Thermoluminescence and Thermoluminescent Dosimetry

Volume III

Editor

Yigal S. Horowitz, Ph.D.

Associate Professor of Physics
Head
Radiation Physics Laboratory
Ben Gurion University of the Negev
Beersheva, Israel

CRC Press
Taylor & Francis Group
Boca Raton London New York

CRC Press is an imprint of the
Taylor & Francis Group, an **informa** business

CRC Press
Taylor & Francis Group
6000 Broken Sound Parkway NW, Suite 300
Boca Raton, FL 33487-2742

Reissued 2019 by CRC Press

© 1984 by Taylor & Francis Group, LLC
CRC Press is an imprint of Taylor & Francis Group, an Informa business

No claim to original U.S. Government works

A Library of Congress record exists under LC control number:

Publisher's Note
The publisher has gone to great lengths to ensure the quality of this reprint but points out that some imperfections in the original copies may be apparent.

Disclaimer
The publisher has made every effort to trace copyright holders and welcomes correspondence from those they have been unable to contact.

ISBN 13: 978-0-367-26261-7 (hbk)
ISBN 13: 978-0-367-26266-2 (pbk)
ISBN 13: 978-0-429-29228-6 (ebk)

Visit the Taylor & Francis Web site at http://www.taylorandfrancis.com and the CRC Press Web site at http://www.crcpress.com

PREFACE

In the nineteen years, 1965-1983, spanning the seven international conferences devoted to luminescence dosimetry, the applications of thermoluminescence to ionizing radiation dosimetry have become legion. There is no doubt that thermoluminescence now occupies a central and constantly expanding role in ionizing radiation dosimetry in the fields of clinical radiation therapy and diagnostics as well as personnel radiation protection and environmental radiation dosimetry.

The commercial development of an extended variety of solid TL dosimeters (TLDs), hot pressed and extruded ribbons, chips and rods, throw-away capsules, and teflon® discs, which can routinely yield 3% precision of measurement, have greatly influenced the acceptance of TLDs in the clinical environment. Loose powders have also remained in use due to the convenience of being able to adjust the dosimeter mass, the slightly higher precision (1 to 2%), and the ability to vary the shape of the dosimeter for specific applications. National and international quality assurance programs using TLDs sent by post are also have an increasingly significant role in establishing widespread confidence in the techniques of TL dosimetry. In addition, working groups have been established for the preparation of international standards for TLDs and TL readers. Reliable TL readers of various levels of sophistication are commercially available, and even the relatively lower priced systems have been found adequate for many routine clinical applications. A parallel development has been the gradually accelerating acceptance of TL dosimetry in the fields of personnel radiation protection and environmental dosimetry. TL dosimetry naturally lends itself to automation and the less stringent requirements for precision are easily fulfilled by TLDs in many radiation environments. A decade of intensive experimental investigation has mapped out the TL properties of many materials and their interaction with various types of radiation fields. Continuing investigations will certainly bring further significant developments. A greater understanding of the difficulties associated with TL dosimetry has also emerged. The TL properties of a particular material are dependent not only on the incorporation of ppm impurities, but also on other defects incorporated into the starting material, or during the material preparation, or by the radiation and thermal treatments applied to the TLD. Thus two materials of nominally identical doping, and even purchased from the same source, may show annoyingly different TL properties. This situation is particularly expressed in the ionization density dependence of the TL efficiency. The nonuniversality of many other important TL properties has led, in certain cases, to significant difficulties of interpretation. For example, it is still not clear to what extent the "over-response" of TLDs to low energy X-rays and the "under-response" to high energy electrons and X-rays are dependent on material composition. Even relative TL properties may also depend on an impressive number of experimental factors. Although several books have been written on TL dosimetry these have been somewhat introductory in nature and too brief to really present to the reader the full complexity of the subject. These volumes are intended to fill the gap both in the depth and breadth of treatment and have thus been primarily written as a reference work for radiation and nuclear scientists and dosimetrists and solid state physicists interested in radiation dosimetry. To this end, effort has been made to include many useful tables, illustrations, and formulas and a comprehensive list of references to the literature is also included following each chapter. Every effort, however, has been made to present the material as logically as possible and with adherence to basic physical principles so that these volumes should also prove useful for TL dosimetrists and scientists just entering the field. Each chapter, covering an important aspect of TL or its application to dosimetry, is written in sufficient detail to give the reader a complete grasp of the subject; the approach is often critical as in the nature of a review article so as to highlight current problems in the field.

The first volume (four chapters) deals with the basic physical and scientific aspects of TL so often essential to optimum use of TL in radiation dosimetry. The first chapter is a general introduction to TL and TL dosimetry as well as to the common quantities used in radiation dosimetry. The definitions follow the recommendations of the 1980 International Committee on Radiation Units. The second and third chapters deal with models of TL trapping and recombination centers and TL kinetics and serve to outline the complexity of the TL mechanisms as well as to present an up-to date discussion of our current understanding of these subjects. The fourth chapter is useful in that it comprehensively lists and discusses the important TL and dosimetric characteristics of the commonly used TLD materials.

The second volume (four chapters) is a hybrid dealing with both TL characteristics (Chapter 4 is a comprehensive discussion of phototransferred TL) and TL dosimetry. In the first chapter, that most important of dosimetric quantities, the TL dose response, is comprehensively reviewed with special emphasis on TL models for supralinearity and sensitization. The second chapter discusses the use of TLDs in various radiation fields (X-rays, beta rays, neutrons, heavy charged particles) and a review and evaluation of general cavity theory is also included. The third chapter introduces the concepts of track structure theory and shows how these can be successfully applied to our understanding of heavy charged particle and neutron TL response.

The third volume (three chapters) is the most applied in nature. The first chapter on TL instrumentation includes many important technical details and is primarily intended to help the TL dosimetrist avoid many of the common pitfalls and errors often encountered in TLD. The second and third chapters treat the clinical and archaeological-geological applications of TLD. These will be especially useful for current workers in the field seeking updated reviews on their specialties.

Yigal S. Horowitz
May 1983

THE EDITOR

Yigal S. Horowitz, Ph.D., is an Associate Professor of Physics and Head of the Radiation Physics Laboratory in the Physics Department of the Ben Gurion University of the Negev, Israel.

Dr. Horowitz received the B.Sc., M.Sc., and Ph.D. degrees in Physics from McGill University in 1961, 1965, and 1968, respectively. He carried out postdoctoral research in experimental nuclear physics at the University of Toronto Linac in 1969 and the Weizmann Institute of Science Tandem-Van de Graaf in 1970. He also carried out medical physics research at the Massachusetts General Hospital and the Massachusetts Institute of Technology in 1971. He joined the Ben Gurion University of the Negev in October 1971 serving as a senior lecturer to 1979 and Head of the Radiation Physics Laboratory from 1975. It was in 1980 that he assumed his present position as Associate Professor of Physics.

Dr. Horowitz is a member of the Canadian Association of Physicists, the Israeli Physical Society, the Israeli Nuclear Society, and the Israeli Association of Medical Physicists. He has been a member of the American Physical Society and the American Nuclear Society and has been the recipient of many research grants from the International Atomic Energy Agency and the United States — Israel Bi-National Science Foundation.

Dr. Horowitz is the author of more than 70 research papers and several review articles in the areas of thermoluminescence and thermoluminescent dosimetry. His current major research interests are in the areas of radiation induced thermoluminescence, radiation damage, track structure theory applied to thermoluminescence, thermoluminescence dosimetry in exotic radiation fields, radiation transport, general cavity theory, and medical radiation physics.

This volume is respectfully dedicated to Professor Herb Attix for his varied and always significant contributions to the field of radiation dosimetry in general and thermoluminescence dosimetry in particular.

ACKNOWLEDGMENTS

I would like to thank Mrs. Helena Paskal of the Ben Gurion University graphic arts department for her heroic efforts in the careful preparation of the over 100 figures used in my own contribution to these volumes. I am also grateful to Professor Shapira, Dean of the Faculty of Natural Sciences, for financial support in the preparation of the manuscript, and to Professor Reuben Tiberger, Chairperson of the Physics Department for his support and for use of the departmental and university facilities.

CONTRIBUTORS

Bengt A. Lindskoug, Ph.D.
Assistant Professor
Departments of Radiation Physics and
  Radiation Therapy
University of Gothenberg
Gothenburg, Sweden

Len-Marie Lundberg, M.S.
Physicist
Radiation Physics Department
Salhgren Hospital
University of Gothenburg
Gothenburg, Sweden

Vagn Mejdahl, Ph.D.
Head
Archaeometry Department
Danish Research Council
Risø National Laboratory
Roskilde, Denmark

Per Spanne, Ph.D.
Department of Radiation Physics
The Medical School
Linköping University
Linköping, Sweden

Ann G. Wintle, Ph.D.
Research Fellow
Godwin Laboratory
Cambridge University
Cambridge, England

THERMOLUMINESCENCE AND THERMOLUMINESCENT DOSIMETRY

Volume I

Introduction
Models of TL Trapping and Recombination Centers
Kinetics of Thermoluminescence Glow Peaks
General Characteristics of TL Materials

Volume II

TL Dose Response
Thermoluminescent Radiation Dosimetry
Track Structure Theory and Applications to Thermoluminescence
Photostimulated Thermoluminescence

Volume III

TL Readout Instrumentation
Clinical Applications of Thermoluminescent Dosimetry
Thermoluminescence Applied to Age Determination in Archaeology and Geology

TABLE OF CONTENTS

Chapter 1

TL READOUT INSTRUMENTATION

Per Spanne

TABLE OF CONTENTS

I. INTRODUCTION

Instrumentation for the study and quantification of the thermoluminescence (TL) from a material is composed of two main parts: a heating unit and a light-measuring unit. The first readout device, used by Sir Robert Boyle to observe TL from diamonds,[1] consisted of his own body heat and the naked eye. Far more sophisticated readout techniques have been developed since then.

The requirements on a TL readout instrument are strongly dependent on the purpose of the TL measurement, e.g., for TL materials research it is essential that the temperature of the sample and that of the heating device closely correspond to enable accurate determination of the sample temperature during readout. This is usually achieved with slow linear heating rates, often less than 1°C/sec. A slow heating rate also usually results in optimum glow peak resolution. On the other hand, for many radiation dosimetry measurements, e.g., routine personnel monitoring of ionizing radiation, it is important that the dosimeter heating is sufficiently rapid to enable the evaluation of many dosimeters in a reasonable time and that the background signals from the readout unit are kept to a minimum.

There exist many reports describing different TL readout instruments designed for various purposes. Some of these are available commercially. In this chapter on TL instrumentation we will, after some introductory remarks on signal-to-noise ratios in TL measurements, treat the principles of different heating systems and light measuring techniques used for investigations of TL materials and TL dosimetry. Finally, some aspects of the auxiliary equipment necessary for reliable TL dosimetry will be discussed.

II. SIGNAL-TO-NOISE RATIOS IN TL MEASUREMENTS

The measurement of TL light emission is always accompanied by systematic and statistical uncertainties. An example of a systematic uncertainty is the error introduced if the dark current of the photomultiplier increases with time during a measuring series because of heating of the photomultiplier photocathode. Statistical uncertainties appear among other reasons because of the quantum nature of light. In specifying the performance of a TL readout system or a TL dosimetry system, it is therefore necessary to specify both the statistical and systematic uncertainties.[2] In the performance, testing, and procedural specifications for thermoluminescence dosimetry in environmental applications, issued by the

American National Standards Institute (ANSI),[3] no distinction is made between the two types of uncertainties, rendering it impossible to evaluate some of the performance requirements.

As a measure of the statistical uncertainty (precision) in a TL measurement, the signal-to-noise ratio defined as $E(S)/\sigma$ or the relative standard deviation which is the inverted value of the signal-to-noise ratio can be used. S stands for the TL signal, $E(S)$ for the expectation value of S, and σ for the standard deviation of the measured TL signal. σ is a measure of the random variations in the measured signal and is often referred to as the noise in the measured signal. In some papers noise has been used to denote the total background signal. This usage should be discouraged since it is the random fluctuations in the signals, not their absolute values that affect the precision in the measurements. A noiseless background signal of whatever magnitude does not constitute a problem since it can be subtracted so that it does not disturb the measurement. In this work noise is used exclusively to denote the random variations in a signal. Whether one uses the signal-to-noise ratio or its inverted value, the relative standard deviation, to specify the statistical performance of a TLD system really does not matter. The only advantage with the signal-to-noise ratio is that it increases as the precision increases.

The TL signal, S, is determined experimentally as the difference between signals in two separate measurements: one measuring the radiation-induced TL plus background signals, the other measuring only the background signals. Sometimes the background signals are measured several times to improve the precision. In this case the signal-to-noise ratio for a measurement can be written as

$$\frac{E(S)}{\sigma} = \frac{E(S)}{\left\{ \sigma_S^2 + \left(\sigma_B^2 + \sigma_{DP}^2 \right) \left(1 + \frac{1}{h} \right) \right\}^{1/2}} \qquad (1)$$

where S is the TL signal; B is all background signals except DP, where DP is the background signal caused by the photoelectric device used for light detection; σ_S, σ_B, and σ_{DP} are the standard deviations of S, B, and DP, respectively; and h is the number of measurements carried out in estimating the total background signal (B + DP).

This equation is valid when S, B, and DP are independent stochastic variables. The factor $(1 + 1/h)$ appears in Equation 1 since the background signals B and DP have to be determined separately under the same conditions as the total signal. If the conditions change between the measurement of the total signal and the background signals, a systematic error is introduced. The signal-to-noise ratio varies slowly with h when h is larger than 5, e.g., it increases by only 5% when h is changed from 5 to 10. This slow dependence on h is due to the fact that Equation 1 is valid when the TL signal is measured with a single dosimeter or a single sample of TL material. It is possible to use several dosimeters in order to increase the signal-to-noise ratio. The optimization problem is then similar to that of dividing a given time interval between signal and background measurements treated in detail by Young.[4]

Analyzing Equation 1 shows that each of its constituents depends on several parameters, some of them interrelated in such a complex way that it is impossible to determine unambiguously the best way for obtaining the optimum performance of a TLD system. However, it turns out that the quantum efficiency of the photoelectric device and the fraction of photons transmitted through the optical filter are the only variables that can selectively decrease the background signals in such a way that the noise in the background signal, σ_B, is decreased by a larger factor than the TL signal, with the net effect that the signal-to-noise ratio increases although the TL signal decreases.[5] A consequence of this is that it is practical to proceed in three steps[5] when optimizing the signal-to-noise ratio for a TLD system, i.e., first optimize the light-collecting and light-measuring equipment exclusive of the optical filter, then the dosimeter sensitivity, and finally the signal-to-noise ratio by reduction and control of background signals.

With respect to the systematic uncertainties caused by the readout equipment, the general guideline to follow in the optimization is to minimize all kinds of drift in the equipment. In this connection one should be aware that drift, e.g., in the light sensitivity of the photoelectric device, certainly affects an experimentally estimated signal-to-noise ratio when the estimate is determined from repeated measurements. However, this does not imply that an estimate of the signal-to-noise ratio alone is sufficient to characterize the performance of a TLD system.

An experimentally estimated signal-to-noise ratio is a meaningful measure of the statistical uncertainty only when the drift in the readout equipment is small compared to the statistical uncertainty.

III. DOSIMETER HEATING SYSTEMS

A. General Aspects

The heating system in a TL readout instrument consists in principle of two parts: a heat supplying medium or device and electronics for the control of the temperature. Two common examples of the former are a resistively heated metal planchet and a jet of hot gas. The temperature control unit, whose general purpose is to produce a reproducible heating cycle (by heating cycle is meant the variation of the temperature with time during readout) may be more or less complicated according to the application in which the TL reader is used. In a simple case it just consists of a transformer that generates an alternating current for ohmic heating of a metal planchet. In more sophisticated readout instruments, a temperature-measuring device is connected to a feedback circuit with programmable reference ramps and plateaux that allows a wide variation of usable heating cycles. By using a curve follower to generate the reference ramps, any heating cycle within the limits of the heat capacity of the heating device can be implemented simply by drawing it on a piece of paper.[5]

One of the most important requirements on a readout instrument for TL measurements is that the heating of the material under study should be reproducible. This is so because for many materials the amount of radiation-induced TL is dependent on the thermal history of the material as well as on the heating rate during readout. Irreproducible heating cycles might therefore affect the noise in the measured TL signal, thereby decreasing the signal-to-noise ratio. In practice reproducible heating is attained by assuring that both the thermal contact between the TL material and the heat supplying device and the temperature cycle of the heat supplying device are reproducible. Which of these two factors imposes most problems in practice is strongly dependent on the components of the heating device and the type of dosimeter. For example, TL dosimeters, consisting of a powdered TL material in a teflon® [polytetrafluoroethylene $(C_2H_4)_n$] matrix,[6] heated with a metal planchet are problematic in the first factor due to their tendency to bend or warp during heating (cf. Section III.B).

The heating of a TL material can in principle be carried out in three different ways. The TL material is placed in thermal contact with a heating medium at a high constant temperature, generally of the order of 300 to 400°C or with a heating medium whose temperature is gradually increased during readout or via radiative heating, e.g., by light pulses. The temperature rise of the TL material is, in the first case, mainly determined by the thermal capacity of the heating device, the thermal conductivity of the TL material, and possibly also by convection. Carter et al.[7] have in fact stated that thermal convection may play the dominant role when powdered TL materials are heated on a metal planchet. In the second case the temperature rise of the heating medium is usually the major factor determining the temperature rise in the TL material, however, one should not underestimate the effect of

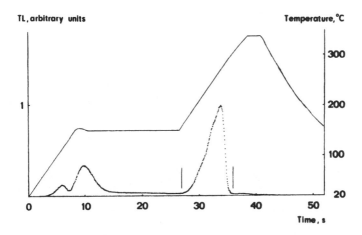

FIGURE 1.   A readout heating cycle with a preheat temperature plateau for fading reduction with TLD-100 (LiF:Mg,Ti) dosimeters. The resulting TL as a function of time is included. The vertical bars indicate the signal integration limits.

the thermal conductivity of the TL material in question. There may exist a considerable thermal gradient through the material, and especially when the heating rate is high, it should not be taken for granted that the temperatures of the heating medium and the TL material are the same. The temperature difference is generally referred to as thermal lag. In this connection it is important to realize that the temperature measured and indicated on the temperature control unit in some TL readers is the temperature of the heating medium and not that of the TL material. Aside from heating cycle reproducibility, a temperature control unit makes possible the use of complex heating cycles, i.e., a heating cycle with a programmable linear heating rate between constant temperature plateaux with programmable temperatures and durations. Such a heating cycle can minimize the influence of room temperature fading by performing a postirradiation anneal in the readout equipment. The postirradiation anneal consists of holding the dosimeter at an appropriately chosen temperature during a fixed time. An example of such a heating cycle together with the TL emission from LiF:Mg,Ti as a function of time is shown in Figure 1. However, one should keep in mind that a postirradiation anneal in the readout unit may be less efficient than one in an external oven.[8] This is so because an annealing period in the readout unit must be short, of the order of 10 to 30 sec, otherwise the total readout time for several dosimeters will be impractically long. When an external oven is used, longer annealing times can be chosen because all dosimeters are then annealed simultaneously.

In LiF:Mg,Ti the cooling rate following readout as well as the rate of heating during readout affects the TL efficiency.[9]

The cooling of the TL material therefore has to be reproducible. The easiest way to cool a dosimeter reproducibly is often to leave it in the readout unit until it has cooled down close to room temperature. This means that one should avoid removing the dosimeter from the readout unit before it has cooled well below 100°C when it is used repeatedly without preirradiation annealing in an oven.

It is well known that nonradiation-induced signals are reduced considerably by removing all oxygen from the readout chamber (cf. Section IV.B.2). In practice this is accomplished by purging an inert gas through the readout chamber during the heating. Nitrogen is often used because it is cheap and easily available. Because of the above-mentioned cooling rate dependence of the TL efficiency and the fact that the cooling rate often depends on the gas flow rate through the readout chamber, it is necessary to control the flow rate.

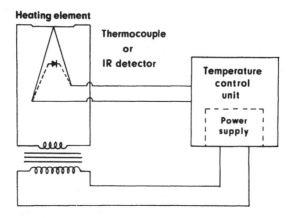

FIGURE 2.   Schematic diagram showing the principle of resistive heating systems with temperature feedback.

B. Ohmic Heating

The most common method by far to heat a TL material for readout is via contact with an ohmic, or resistive, heating element. Such a heating element consists of a piece of metal with low resistance, of the order of some milliohms, which is heated by passing an alternating current through it. The alternating current is obtained by connecting the heating element to the secondary winding of a low voltage transformer. The current used to generate a heating rate of the order of 10°C/sec is often several amperes. With this heating system the heating cycle can be controlled simply by varying the voltage over the primary winding of the transformer.

In TL readers with resistive heating elements the dosimeters can be either directly heated by direct contact with the heating element or indirectly heated by placing them on a metal tray or cup in contact with the heating element. The latter method has the disadvantage that one has to assure reproducible heat transfer between three bodies while for the former only two are involved. On the other hand, indirect heating makes possible a permanent electrical connection between the secondary winding of the transformer and the heating element while at the same time allowing easy removal of the dosimeter holder, e.g., a Pt cup, for cleaning or exchange. The permanent electrical connection enables the achievement of a reproducible heating cycle because the possibility of varying contact resistance between the connection leads and the heating element is eliminated.

The resistive heating elements that are used in TL readers can be divided into two main categories of variable and constant temperature during readout.

1. Heating Elements with Varying Temperature

In early TL readers dosimeters were often heated on a metal planchet, usually made of Kanthal®, through which a stabilized alternating current was fed for a fixed time. This heating system does not result in linear heating but in a kind of sigmoid temperature rise vs. time. The heating rate can, however, be varied to some extent by selecting the magnitude of the heating current. The maximum temperature can be adjusted by varying the duration time of the alternating current.

Since the maximum readout temperature depends on the temperature when the heating current is turned on, it is important that the heating element cools down to the same temperature after each heating cycle. If this point is considered such a simple heating cycle may be sufficient for many routine TLD measurements. Carlsson et al.[10] have reported that a relative standard deviation as low as 0.2% can be obtained in measurements with LiF-teflon® dosimeter readouts in an instrument utilizing this heating principle.

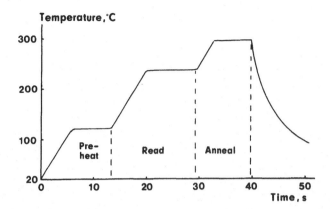

FIGURE 3.   Readout heating cycle with a read temperature plateau to minimize the nonradiation-induced signals as suggested by Webb and Phykitt.[13]

One of the main advantages of resistive heating is that it is a relatively simple task to design a temperature control system with preselected heating cycles. In such a temperature control system (cf. Figure 2), the signal from a temperature-measuring device is compared with a reference ramp and the heating current varied according to the difference between them. In most cases a thermocouple is used to provide the negative feedback signal to the electronics controlling the heating current. In order to obtain a reproducible heating cycle it is, of course, essential that the thermal contact between the thermocouple and the heating element is reproducible. The best method is to weld the thermocouple to the heating element, but this makes its removal impossible, which might be a problem if the dosimeters are directly heated, e.g., with a planchet which tarnishes in repeated use. Pressing the thermocouple mechanically against a planchet is often used but may give rather irreproducible heating cycles.

One method that overcomes the above problem is to use infrared sensors to measure the temperature of the heating device. This is utilized in a commercially available instrument (Teledyne® Isotopes TLD 7300 Reader) with a planchet heating system. In this readout instrument the dosimeter is placed on a metal planchet which, with the aid of a drawer assembly, is slided into position between electrical contacts. The infrared sensor measures the infrared emission from the black painted backside of the planchet. The major obstacle that may be experienced with this temperature-monitoring method is that dust may change the light path between the planchet and the infrared sensor.

The design principles and problems in connection with the construction of a feedback system for the control of the heating cycle in a TL reader have been treated in considerable detail by Robertson.[11]

One disadvantage with resistive heating, direct or indirect, is that incandescent light emission from the heating element may contribute significantly to the noise in low dose measurements (cf. Section IV.B.1). The number of photons emitted per unit time due to incandescence depends on the area of the heating element and varies with the temperature. In order to estimate the magnitude of the PM signal caused by incandescent light emission from a metal, Burch[12] calculated the emission rate of a blackbody radiator as a function of the wavelength for different temperatures in the interval of 100 to 400°C. He found that blackbody emission at 300°C can be neglected for wavelengths below about 600 nm (cf. Section IV.B.1). In many TL measurements the problems arising from incandescent light emission can be solved by utilizing an optical filter with small transmission above 600 nm.

With a variable temperature heating element the incandescent light emission is observed as a high temperature tail in the glow curve. The TL signal integration in absorbed dose measurements should therefore be terminated at as low a temperature as the TL glow peaks allow in order to minimize the contribution of the incandescent emission to the noise.

In order to minimize the noise due to incandescent light emission, Webb and Phykitt[13] suggested the use of a readout cycle with a "read zone" in which the temperature of the heating element is rapidly raised from the preheat temperature to a constant temperature chosen to empty the TL peaks during the read zone (Figure 3). With this type of heating cycle a lower temperature can be used during signal integration than when a linear heating rate is used. This is so because the temperature of the glow peak maximum increases with increasing heating rate.[14] However, as the detected radiation from a 1 cm² blackbody radiator at 300°C is negligible below 600 nm,[12] this method to reduce incandescent light emission is most valuable for glow peaks above 300°C or for TL materials with emission spectra in the red wavelength region. In other cases the use of an optical filter might be equally efficient. In fact, there does not exist any reports on quantitative evidence that a read zone as described above really improves the signal-to-noise ratio in low dose measurements.

One of the main advantages with resistive heating is that the heating element can be designed to fit many different shapes of dosimeters. In instruments with indirect heating one can easily exchange different dosimeter holders so that many different types of dosimeters can be read out with the same instrument. Special care has to be taken, however, when designing a heating element for teflon® dosimeters because of their tendency to bend during heating, thereby changing the contact with the heating element. There are two alternative ways of preventing disc-shaped teflon® dosimeters from bending during heating. One is to press the heating element with the dosimeter on top of it onto a transparent window, e.g., made of quartz. A heating element in the shape of a mesh may then be placed between the dosimeter and the quartz window thereby decreasing the thermal gradient in the dosimeter.[15] As the window is heated during readout, one has to be careful that its own thermoluminescence does not disturb the measurements[16] (cf. Section IV.C). The second alternative is to place the teflon® dosimeters under a mesh or sprong welded on top of a planchet. A slight disadvantage with the mesh is that it prevents some of the light from reaching the PM tube. An advantage is that the dosimeter is heated from both sides.

2. Heating Elements with Constant Temperature

Many routine TLD measurements do not require a programmable heating cycle or a linear heating rate. In such measurements a heating element consisting of a metal block at a high constant temperature can be used. The dosimeter is heated by contact with the heating element.[17-20] Such a heating device is often referred to as a hot finger or hot anvil. If the thermal capacity of the heating device is high enough, the heating cycle is mainly determined by the thermal conductivity of the dosimeter. When a hot finger is brought in contact with a dosimeter, the dosimeter temperature first increases rapidly and thereafter asymptotically approaches a constant value. The thermal gradient in the dosimeter is large during the initial part of the heating cycle because of the high initial heating rate but is reduced during the later part. In one instrument utilizing a hot anvil a thermocouple is pressed against the dosimeter on the side opposite to the hot anvil and the heating terminated when the temperature on the back side of the dosimeter is high enough to assure that the TL emission is complete. If the heating takes too long an alarm is sounded and the reader stops.[18,22] In this way one can be confident that the thermal gradient in the dosimeter does not cause incomplete readout of the TL.

Although hot fingers are used mainly for direct heating, they can be used for indirect heating as well. The dosimeter material must then be fixed in a holder so that the heating element can be pressed firmly against it. In practice this is achieved by placing the dosimeter between two transparent heat-resistant foils mounted in a hole in a metal card[19,20] and by sticking the dosimeter tablet to an adhesive tape also mounted in a hole in a metal card. Another possibility would be to use a clamping ring. A disadvantage with all these methods is that the materials used to hold the dosimeters are also heated during readout. This may cause emission of nonradiation-induced TL.[21,22] Another problem that may appear when the dosimeters are placed between two foils is that their transmission may change in repeated use.[22] Furthermore, when the foils are made of teflon®, thermal treatments at temperatures above approximately 320°C are impossible without dismounting the dosimeter holder because teflon® softens at 327°C.

The most obvious advantages with the hot finger heating method are its simplicity and rapidity. This has led to its incorporation in automated instruments developed for routine personnel monitoring which require large dosimeter throughput in a limited time.[18-22]

The heat dissipation from a heating element continuously kept at 300°C is rather high and it is almost mandatory to cool the readout chamber and the PM tube. Another disadvantage with a hot finger as compared to heating elements with varying temperature is that the PM signal caused by incandescent light emission from the heating element is present during the whole readout cycle.

The use of hot fingers in TL readers does not exclude the possibility of carrying out a preheat in the reader (cf. Figure 1), but two elements must then be used, one for the preheat and one at a higher temperature for the readout of the dosimetry peaks.[23]

C. Hot Gas

In 1968 Petrock and Jones[24] described a TL reader in which a jet of hot nitrogen of constant temperature was used to heat TL dosimeters. The dosimeters were lifted into the nitrogen jet in the readout chamber on top of a suction needle pneumatically raised from below the dosimeter through a hole in the sample changer. Two types of sample changers were described, one manual in the shape of a drawer assembly and one in the shape of a turntable, which allowed automated readout of up to 80 dosimeters. More recently, readout instruments have been developed utilizing hot gas heating either with a nitrogen jet[25-32] or with an air jet.[33,34]

Although the heating principle is the same for all these instruments, they differ in several important respects as will be described in the following.

Heating with a jet of hot gas has several advantages over other heating methods: the heat transfer from the gas to the dosimeter is very efficient and results in rapid readout, incandescent light emission is greatly reduced, oxygen-induced background signals are easily suppressed by using an inert gas as heating medium, and for some dosimeter materials nonradiation-induced signals caused by surface effects can be excluded from the measurements by a simple method. All these advantages are most pronounced for instruments in which the dosimeters are lifted out of the sample changer into the gas jet.[24,25,27-29]

That the most efficient heat transfer is obtained when the dosimeter is lifted out of its holder or sample changer is evident from the fact that the heating medium then almost completely surrounds the TL material. With a nitrogen flow rate of 5 ℓ/min and a gas temperature of the order of 350°C, a heating time of 10 to 15 sec is typically necessary for readout of dosimeter pellets consisting of 40 mg LiF:Mg,Ti. In order to read out dosimeters with large surface areas, like teflon® discs, Julius et al.[29] incorporated three gas jets blowing on the dosimeter from different directions in the same plane. It is difficult to anticipate if the heat transfer efficiency in this technique is also increased for smaller dosimeters because of turbulence due to crossing of the jets.

Incandescent light emission is minimized if the dosimeter is lifted out of the sample changer into the hot gas jet because apart from the TL material very little material is heated to a temperature where incandescent light emission contributes significantly to the noise. It is advisable to direct the gas jet perpendicular to the light path from the TL sample to the

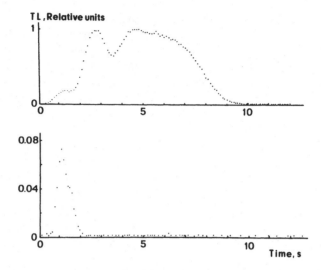

FIGURE 4.   TL as a function of time for TLD-700 dosimeters non-
linearly heated with a 370°C hot nitrogen jet. Top: TL induced by ^{60}Co
γ-irradiation to 6 mGy. Bottom: Nonradiation-induced signal.

PM tube. In this way the PM tube sees a minimal amount of heated material. This geometry is impossible to obtain unless the dosimeter is lifted out of the sample changer.

In the readers designed by Petrock and Jones,[24] Bøtter-Jensen,[25-28] and Julius,[29,30] gas outlets in line with the gas jets were included in the walls of the readout chambers. Although this prevents excessive heating of the readout chamber so as to minimize incandescent light emission, the heat dissipation is high which makes it almost mandatory to cool the readout chamber or the PM tube or both to avoid heating of the PM photocathode. In this connection it is important to remember that many TL readout instruments with PM tube cooling are equipped with a cooling system that does not monitor or actively control by feedback the temperature of the photocathode even though this is the critical point to keep at a constant temperature. It is therefore often necessary in readers with hot gas jets, with considerable heat dissipation in the readout chambers, to delay the readout of the first dosimeter to a time sufficiently long after turn on of the gas flow to allow the PM photocathode to reach thermal equilibrium with its surrounding. The delay time is dependent on the thermal in-sulation between the readout chamber and the photocathode and therefore depends on the type of readout unit. The delay time can easily be determined by monitoring the dark current from the PM tube as a function of time after turn on of the gas flow. The importance of the delay was clearly illustrated by measurements in our laboratory of the photocathode temperature in one instrument with indirect cooling of the PM tube. The cooling was performed by circulating a liquid at 1°C around the PM tube housing. Despite the low temperature of the PM tube housing, the PM tube photocathode reached an equilibrium temperature of 12°C about 40 min after the turn on of the gas flow (5 ℓ/min, 370°C).

By using an inert gas like nitrogen as heating medium, oxygen-induced TL signals are automatically suppressed. The suppression is most efficient when the dosimeter is lifted out of the sample holder because the inert gas then almost completely surrounds the TL material. Also nonradiation-induced TL from parts other than the TL material are minimized. It has been pointed out that the focusing of the gas jet is critical in this respect when the dosimeter is heated without lifting it out of the holder.[32] If the gas jet is too broad the sample holder is heated and may emit nonradiation-induced TL.

Even if an inert gas is used for heating of TL dosimeters several types of nonradiation-induced signals may still be present which are usually ascribed to surface effects. For

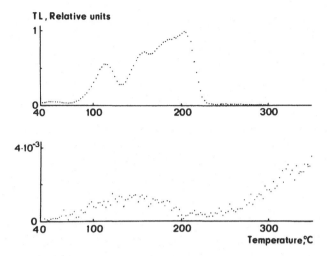

FIGURE 5.   TL as a function of time for TLD-700 dosimeters heated
with an approximately linear rate (15°C/sec). Top: TL induced by ^{60}Co
γ-irradiation to 6 mGy. Bottom: Nonradiation-induced signal.

LiF:Mg,Ti dosimeters these may be reduced to a level equivalent to a radiation-induced TL
signal of 0.5 μGy by a chemical treatment of the dosimeter surface.[5] Recently nonradiation
induced, or spurious, signals appearing during the first seconds of the readout have been
observed when LiF:Mg,Ti and Li$_2$B$_4$O$_7$:Mn dosimeters were heated with a jet of hot nitrogen[35,36]
(cf. Figure 4). These spurious signals caused considerable trouble because they were in some
cases equivalent to an absorbed dose of 300 μGy. All dosimeters did not emit such a signal
and the estimated relative standard deviation for the background signals for a series of 15
dosimeters was in several cases as high as 100%. However, starting the signal integration
2.5 sec after the dosimeter pellet had been placed in the nitrogen jet completely eliminated
the noise contribution from the spurious signals. This delay increased the signal-to-noise
ratio in low dose measurements by a factor which varied between 6 and 60 for different
measuring series on 15 dosimeters.

At first sight the spurious signals seem to be of an hitherto unknown kind because they
appear to have a readout temperature of only 60°C when compared to the radiation-induced
TL (cf. Figure 4). The spurious signals emitted from LiF:Mg,Ti dosimeters read out with
a planchet heating system using a linear heating rate are observed at considerably higher
temperatures (cf. Figure 5). However, Spanne[36] has shown that the spurious signals caused
by hot gas had a readout temperature between 100 and 200°C and were caused by surface
effects. The rapid emission can then be explained by the heating technique. When a dosimeter
pellet is placed in the hot nitrogen jet its surface is heated much more rapidly than its interior.
Signals caused by surface effects therefore appear earlier during readout than the radiation-
induced TL with corresponding readout temperatures. It therefore seems likely that the
spurious signals observed for a planchet temperature between 100 and 200°C (N$_2$:0.5 ℓ/min)
are the same as those observed with the hot nitrogen system although the latter appear very
early during readout. Also the spurious peak with maximum above 300°C may contribute
to the early spurious peak during dosimeter heating with 370°C hot nitrogen.

The very rapid emission of the spurious signals during heating with a jet of hot nitrogen
is an advantage compared to other heating systems because it makes it possible to discriminate
between spurious and radiation-induced signals although they have the same readout temperature. With a 2.5-sec delay of the start of the signal integration interval it is possible
with LiF:Mg,Ti dosimeters to measure absorbed doses down to at least 3 μGy if the low
detection limit is defined as the absorbed dose which results in a signal-to-noise ratio equal
to 3.[36]

A disadvantage with the hot gas heating method is that it is very difficult to design an instrument with programmable heating rates and plateaux. Bøtter-Jensen[27] partly solved this problem by constructing a reader with two independent gas jets with different temperatures. Preheating of the dosimeter is performed in the low temperature jet, TL signal readout in the high temperature jet. Another disadvantage with hot gas heating is that powdered TL dosimeters cannot be read out unless they are kept in a transparent container, e.g., a glass tube.

Because of the previously mentioned advantages, TL readers utilizing a hot gas jet as heating medium are particularly well suited for low dose measurements. They have therefore found application in personnel and environmental monitoring of ionizing radiation. Several instruments with automatic sample changers have been developed especially for these purposes.

D. Optical Heating

Cameron et al.[37] described a simple design for a TLD reader with optical heating. In this reader the dosimeters were placed on a stainless-steel planchet which was heated by focusing a 150-W projection lamp on its back side. The lamp was switched on (off) at the start (stop) of the readout. A light "dimmer" in the lamp circuit was used to control the light flux and thus the heating rate. The main problem with this reader was keeping stray light from reaching the PM tube. A TL instrument with similar design has been used in archaeological TL dating.[38]

Recently a TLD system utilizing optical heating has been described.[39] Because of the infrared (IR) heating system only specially designed dosimeters can be used. The TL material in the dosimeters is in powdered form and placed as a monolayer on a reflective plate with a black backside. The powder is kept in place by a transparent plastic foil (20 mg/cm²). The TL material is heated by flashing IR radiation from a pulsed tungsten lamp on the back side of the TL element. The manufacturer claims that readout can take place in as short a time as 0.8 sec.

An evident advantage with the IR heating method as implemented in this system is the rapid readout. However, the rapid readout is possible only with thin dosimeters of small mass. This advantage therefore is compromised by a relatively small TL signal resulting in a low dosimeter sensitivity (TL detected per unit energy imparted). This is especially troublesome when nearly tissue-equivalent TL materials, e.g., $Li_2B_4O_7$:Cu, are used because their TL efficiencies are much smaller than for some high Z materials, e.g., $CaSO_4$:Dy. The thin dosimeters are, however, an advantage when measuring absorbed doses from low energy beta radiation.

A rapid readout of the order of 1 sec severely restricts the number of TL materials that can be used. Only TL materials that do not exhibit thermal quenching can be read out because of the extreme heating rate. TL materials like LiF:Mg,Ti and CaF_2:Mn are therefore poor candidates for rapid readout using IR heating (cf. Gorbics et al.[14]).

E. RF Heating

Brunskill and Langmead[40] have reported on the use of radiofrequency (RF) heating of TL dosimeters. The TL material was bound to a graphite tray and heated by the current induced in the graphite by a flat RF induction coil situated just below the dosimeter. With this heating system it was possible to raise the graphite tray from room temperature to 300°C in about 10 sec.

Since the energy is radiatively transferred to the dosimeter in RF heating systems, it is possible to bind TL material to both sides of the graphite tray. This makes it possible to measure both surface and penetrating doses with the same dosimeter if the TL is measured with two PM tubes facing the dosimeter from opposite sides.

RF heating has proved reliable but has not found widespread use. McKinlay[1] assumes that this is the case because of the impossibility of direct temperature measurement necessary for complex heating cycles. However, temperature measurements utilizing monitoring of IR radiation on the back side of the graphite tray carrying the TL material should be a possible remedy to this problem.

F. Internally Heated Dosimeters

During the early 1960s much of the TL work took place with powdered TL materials. This caused considerable trouble because nonradiation-induced signals caused by surface effects seriously affected the signal-to-noise ratio in low dose measurements. Furthermore, powdered TL materials are difficult to handle. Dosimeters consisting of a TL material and an ohmic heating element encapsulated in a glass bulb were therefore developed.[41] Examples of the design of such internally heated dosimeters are TL powder bound on an ohmically heated carbon plate, a metal cylinder coated with a TL phosphor and heated with an ohmic filament, or a coaxial platinum filament surrounded by TL material. All heating elements in internally heated dosimeters are ohmic in nature and they differ from other ohmic heating methods only by the fact that the heating device is incorporated into the dosimeter.

Bulb dosimeters are a very convenient way to handle TL powder. An additional advantage is that spurious signals due to oxygen are also minimized by filling the bulbs with inert gas. Serious drawbacks are that the dosimeters are rather large and the dosimetric interpretation of the signals is rather complex due to the glass wall and the metal heating element which are both present during the irradiation. Furthermore, in low dose measurements it is necessary to correct for the build-up of TL signal with time caused by beta particles from ^{40}K in the glass.

G. String Dosimeter Heating

In 1967 Benner et al.[41] and Lindskoug et al.[43] described a TL dosimetry system with a string dosimeter consisting of a TL material enclosed in a narrow ($\emptyset$ = 1.5 mm) teflon® tube. The dosimeters were heated by inserting the tube along the central axis of a brass heating coil through which a current of 5 A was fed during readout. Although the heating of the coil is ohmic in nature, this heating system differs from other ohmic heating systems because the dosimeter is not in direct contact with a solid heating element. An automated readout unit using the same heating system was later developed.[44,45] In this readout unit the dosimeters, which consisted of TL material separated by teflon® spacings enclosed in teflon tubes, were heated by continuously feeding the string dosimeter through a glass tube situated inside the heating coil. The temperature in the glass tube was monitored with a resistive temperature sensor and kept constant by a feedback system regulating the current through the coil. With this heating system glow peaks with different readout temperatures cannot be resolved since the temperature of the string dosimeter is higher at the entrance than at the exit of the readout oven. TL with different readout temperatures is therefore emitted simultaneously. To eliminate fading errors due to low temperature glow peaks the dosimeter strings were fed through an extra heating coil, used as a preheater oven, before entering the readout chamber. This extra coil was kept at a temperature that only caused the low temperature TL to be emitted.

In order to improve the signal-to-noise ratio Lindskoug[46] suggested the use of teflon® tubes filled with TL material over their whole length so that the TL signal could be integrated over an arbitrary length. Alternatively one can sample the TL signal repeatedly during the translation of the dosimeter through the readout oven. In principle one can then measure the dose distribution along the whole dosimeter length. This technique is of considerable interest in radiation therapy dosimetry measurements, e.g., in intracavitary dose measurements (cf. Volume III, Chapter 2) and therefore is in widespread use in radiation therapy departments. However, there is one serious disadvantage associated with LiF-teflon® rods in tubes which is common to all kinds of LiF-teflon® dosimeters used to measure absorbed doses of the order of 1 Gy or larger irrespective of the readout principle. Teflon® cannot withstand temperatures above about 325°C and therefore cannot be subject to any 400°C annealing. This means that if the dosimeters are used repeatedly at dose levels of the order of 1 Gy or larger they will become sensitized. Lindskoug calibrates all dosimeters individually every time he uses them to take account of this effect. However, when the sensitization increases with increasing accumulated absorbed dose the LET dependence also changes.[47,48] This means that it is necessary to use radiation with the same LET when assessing the sensitivity changes as during the dose measurements, otherwise systematic errors are introduced. The LET dependence therefore makes it difficult to use teflon® dosimeters in mixed radiation fields when the absorbed doses are at a level where sensitization occurs.

IV. LIGHT COLLECTION UNIT

The light collection unit in a TL readout system generally consists of three parts: a readout chamber in which the TL is generated by heating, an optical window, and a light guide system between the readout chamber and the photoelectric device used for light detection. The optical window, generally an optical filter, together with a space between the TL material and the window serves as thermal insulation between the heating medium and the light detector. The thermal insulation is necessary because photoelectric light detectors are sensitive to temperature changes.

A. Readout Chamber

Theoretically, the best design of a readout chamber is a parabolic mirror with the TL material placed at the focus. All light reflected by the mirror then leaves the readout chamber as a parallel beam and only the light escaping directly through the opening of the mirror is divergent. The parabolic mirror therefore increases the light reaching the light detector. If the parallel reflected light is perpendicularly incident on the optical filter the loss of light due to reflections at the interfaces of the optical filter is minimized. However, in practice it is difficult to implement this construction principle fully, partly because of difficulties in placing the TL material at the focus. It is possible in readers where the dosimeters are lifted out of the sample changer and heated with a jet of hot gas. However, in most cases one has to be content with simpler solutions than a parabolic readout chamber.

A TL instrument has been described in which the favorable properties of a parabolic readout chamber have been utilized.[49] In this instrument which is designed for readout of small amounts of TL material, half a parabolic mirror is used to collect the TL on the PM tube. In principle the mirror is obtained by cutting a parabolic mirror in a plane through the rotation axis of the parabola generating it. The mirror is placed above the heating element with the TLD at the focus and the surface of the heating element in the cutting plane. In principle 4π light collection is obtained in this way if the heating element is reflective. In practice the light collection is not ideal because the TL material has a finite extension and is generally not completely transparent to its own light, which causes some light reflected in the heating element to be attenuated.

It has been debated whether it really is an advantage to make the walls of the readout chamber reflective. There are two main reasons for questioning reflective readout chambers. One is that small amounts of dirt are released from the dosimeters during heating and this changes the reflectivity, the other is that when the readout chamber is reflective small changes in the position of the TL material can cause larger changes in the light collection efficiency than when it is not reflective. The first argument is evident from the experience of Grogan et al.[50] and the second from the work of Karzmark et al.[51] Karzmark et al. showed that positioning errors of the order of 1 mm may cause light collection efficiency variations of the order of several percent. The light collection efficiency variation is most pronounced for displacements perpendicular to the photocathode surface and when the photocathode-to-dosimeter distance is small.[51] In practice it is often possible to reposition a dosimeter in the readout chamber within less than 1 mm, but positioning variations might still cause significant variations in light collection efficiency.[51] When one decides whether or not to use reflective walls in the readout chamber one, therefore, has to consider the application in which the reader is to be used and the connected uncertainties. The uncertainties due to positioning variations are generally small compared to the noise from other sources when low dose measurements are performed with TL dosimeters. In this case reflective readout chamber walls are to be preferred. On the other hand, when the TL signal is large one might in some cases optimize the precision by using nonreflecting walls.

In commercially available TL readout equipment, it is generally difficult to alter the readout chamber so as to achieve more efficient light collection. However, some methods do exist to increase the signal-to-noise ratio and decrease the systematic uncertainties.

In readout equipment where the TL material is placed on a reflective heating medium like a metal planchet or cup during readout, the reflectivity may change in repeated use due to tarnishing.[50,52,54] The reflectivity may in some cases be retained by keeping the planchet or cup in H_2O_2 overnight. This may also eliminate nonradiation-induced TL signals from a Kanthal® planchet observed after long use and ascribed to oxide and dirt accumulated on the surface.[55] In this respect platinum (Pt) planchet should be used, if possible, because Pt does not oxidize in air and is highly reflective.

Since small amounts of dirt may be released from the TL material during heating,[50] the whole of the readout chamber inclusive of the optical filter should be cleaned at regular intervals. The cleaning of the readout chamber is especially important when teflon®-based dosimeters are used[6] because PTFE softens at temperatures above 300°C and small PTFE molecules then evaporate and cause a sticky film to adhere on the walls of the chamber.[56]

Reference light sources are sometimes used in an attempt to correct for the effects of build-up of dirt in the readout chamber. However, the reference light sources seldom have the same emission spectrum as the TL material under study and never have the same geometric shape as the dosimeters. This means that they cannot fully assess a change in light collection efficiency. Therefore it is often better to use dosimeters irradiated to a known absorbed dose in a standard fashion and treated in exactly the same way as the other dosimeters than to use a reference light source. With such reference dosimeters, the sensitivity of the whole TLD system is determined.

B. Optical Filters

Optical glass filters are used in most TL read units to isolate the housing of the light detector thermally from the readout chamber. The isolation is obtained by preventing heat convection to the light detector and by reflection and absorption of IR radiation. Commercial readers are generally equipped with filters that reflect and absorb photons in the IR wavelength region but with high transmission in the whole visible region. This makes it possible to read out dosimeters with different emission spectra with the same readout unit. However, if only one TL material is used the signal-to-noise ratio can often be increased significantly by an optical filter with a more narrow band pass, selected especially to suit the material that is used. The increase is due to reduction of the noise in the detected incandescent light and the nonradiation-induced signals from the dosimeters.

When choosing optical filters for a TL application it is important to realize that the signal-to-noise ratio sometimes may increase although the number of TL photons impinging on the light detector decreases. This follows from Equation 1.

1. Incandescent Light Emission

When a TL material is heated during readout, both the TL material and the readout

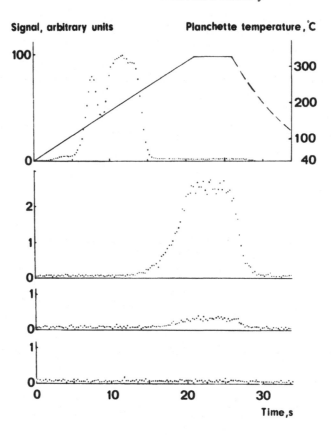

FIGURE 6. The upper curves show the readout heating cycle and the corresponding γ-radiation-induced TL from a TLD-100 dosimeter. The curves below show the nonradiation-induced signals measured with different filter combinations and without any dosimeter. From the top to the bottom: IR rejecting filter supplied by the TLD readout equipment manufacturer + 3-mm thick OG3 filter, IR + OG3 + 4-mm thick BG3, and IR + OG3 + 3-mm thick BG12 filter.

chamber emit radiation of the same origin as blackbody radiation. This radiation is often referred to as blackbody radiation, although neither the TL material nor the readout chamber is a blackbody. Other commonly used names are IR radiation and incandescent light. Since the radiation might well be in the visible wavelength region, incandescent light emission is used in this text.

The number of incandescent light photons emitted from a material depends both on its area and its temperature and varies with the wavelength. In order to estimate the effects of incandescent light emission on TL measurements Burch,[12] calculated the emission of a blackbody radiator as a function of wavelength for different temperatures in the interval of 100 to 400°C. He found that for a 1 cm² blackbody radiator at 300°C about 120 photons per second are detected by a S-11 (Cs_3SbO) photocathode if its diameter subtends an angle of 60° at the radiator. Blackbody emission below 600 nm makes a negligible contribution to the total number of detected blackbody photons at 300°C. The problems arising from incandescent light emission are therefore not of major importance if an optical filter with a cut-off for wavelengths longer than 600 nm is used in measurements where the signal integration takes place below 300°C.

Burch[12] pointed out that for a blackbody radiator at 300°C the number of emitted photons per unit wavelength interval increases so rapidly with increasing wavelength that the number

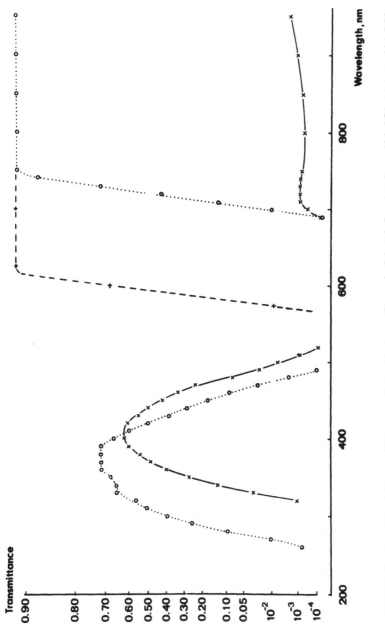

FIGURE 7.   Optical transmittance of filters manufactured by Jenaer Glaswerke Schott & Gen. (+) 3-mm thick OG3. (o) 4-mm thick BG3. (×) 3-mm thick BG12.

of photons giving rise to pulses at a PM tube anode may be significant even for wavelengths as long as 900 nm. The increase in the number of emitted photons per unit wavelength interval then more than compensates the decrease of the PM photocathode quantum efficiency with increasing wavelength. For Cs_3Sb photocathode Burch[12] arrived at a figure of a few detected photons per second in a 10-nm interval around 900 nm at 300°C. Since the black-body emission increases rapidly with temperature, it is essential, at least in connection with some especially red-sensitive PM photocathode materials, to make sure that the optical filter is thick enough for photons with wavelengths up to 900 nm if TL signals at temperatures larger than 300°C are to be measured. High temperature TL measurements are frequent in TL dating.

The importance of considering filter transmissions above 700 nm in TL measurements is illustrated in Figure 6. This figure shows the nonradiation-induced signals that were detected with different filter combinations in a TLD reader utilizing a planchet heating system. In all the measurements in Figure 6 an IR-absorbing filter supplied by the TLD readout equipment manufacturer and a Schott OG3 filter were used. The OG3 filter only transmits light with wavelengths longer than 600 nm. By adding filters with negligible transmission between 600 and 700 nm and different transmissions above 700 nm, the importance of reducing the incandescent light emission above 700 nm could be evaluated (Figure 6). An EMI 6097S PM tube was used which is not especially red sensitive. Transmission curves for the OG3 and the added filters are shown in Figure 7. For comparison, the heating cycle and the corresponding glow curve from a LiF:Mg,Ti dosimeter (TLD-100) are included in Figure 6.

Taking the filter transmissions into consideration, Figure 6 shows that although the IR absorbing filter was used, it was possible to detect nonradiation-induced signals from the readout chamber with wavelengths longer than 700 nm only if the temperature was high enough. This signal could be efficiently reduced when a BG12 filter was used. The back-ground signal from the readout chamber was then negligible in comparison with the dark current. This illustrates the necessity to consider filter transmissions not only in the visible wavelength region, but also for longer wavelengths.

The temperature dependence of the incandescent light emission often makes the corresponding signal appear as a high temperature tail in TL measurements as is the case in Figure 6. The TL signal integration interval should therefore be terminated at as low a temperature as the TL emission allows.

Practical situations where incandescent light with wavelengths longer than 700 nm may contribute significantly to the noise are in TL dating and when the sixth glow peak from LiF:Mg,Ti is used to estimate the absorbed dose from neutrons.[57]

2. Reduction of Nonradiation-Induced TL

The spectral distributions of the radiation- and nonradiation-induced TL differ for some materials. This makes it possible to selectively decrease the noise due to the nonradiation-induced signals by the use of optical filters. A consequence of the selective suppression is that when an optical filter is introduced the signal-to-noise ratio may increase although the detected radiation-induced TL decreases.

The spectral distributions of the nonradiation-induced TL have been studied for both LiF:Mg,Ti[59-62] and CaF_2:Mn[59] readout in air. All of these measurements show that the emission spectrum of the radiation-induced TL differs from that of nonradiation-induced TL. For LiF:Mg,Ti the latter emission is predominantly in the red wavelength region, while for the former it is in the blue region. Considerable improvements in the signal-to-noise ratio can therefore be obtained by using an appropriate optical filter when LiF:Mg,Ti is read out in air. The improvement is smaller for CaF_2:Mn because the difference between the emission spectra is smaller than for LiF:Mg,Ti. The choice of filter is determined not only by its transmission properties in the visible region, but also in the IR wavelength region as pointed out in Section IV.B.

Emission spectra of nonradiation-induced TL from LiF:Mg,Ti have also been measured during readout in O_2, N_2, and Ar. It was found that the emission spectrum changes when the gas in the readout chamber is changed. These spectral changes appear because the removal of oxygen from the readout chamber suppresses one component of the nonradiation-induced TL so that the total nonradiation-induced signal is reduced by a factor between 20 and 50.[51,62,65]

Other materials also profit from readout in an oxygen-free atmosphere.[59,64,66] Removal of oxygen from the readout chamber during readout does not suppress all types of nonradiation-induced signals caused by surface effects and it is worthwhile to use optical filters even if an inert gas is used during readout. Furthermore, an optical filter may make the measurements less sensitive to small variations in oxygen impurity of the gas in the readout chamber.

C. Light Collection on the Photoelectric Detector

When it comes to optimizing the light collection after the window to the readout chamber, the principle is simple: the light collection should be maximized without introducing any new background signals. The solid angle that the light detector subtends at the readout chamber is maximized if the light detector is placed as close to the readout chamber as possible. However, to avoid gradual heating of the light detector it is often advisable to keep it some distance away from the window to the readout chamber. One widely used method to increase the light collection efficiency is then to use a light guide between the window and the light detector. Another method to minimize heating of the light detector is to bend the light path with a mirror so that the light detector does not have to be placed above the readout chamber.[29,67] This minimizes heating due to convection. However, when designing light guides several precautions must be taken to avoid introducing noise due to new background signals. Solid light guides, e.g., made of quartz glass, have often been placed between the readout chamber and the PM photocathodes in TL readers. Not unusual are 15-mm thick glass light guides. However, solid light guides should be avoided because environmental radiation may induce scintillations, phosphorescence, and Čerenkov light in them. The resulting light pulses may be large and cause significant noise. Instead of solid light guides, tubes with reflective inner surfaces should be used together with thin heat-absorbing filters.[5] However, if solid light guides are used together with optical filters it is advisable to place the filter between the light detector and the light guide instead of in front of the light guide because it then attenuates some of the light produced in the light guide.

Another noise source that may be introduced by a light guide is the disturbance of the PM dark current due to an object not at cathode potential placed too close to the cathode (cf. Section V.C.4). This means that when a PM tube photocathode is biased at a negative potential relative to ground it is necessary to terminate the light guide at a sufficiently long distance from the photocathode to avoid increasing the dark current. Any increase in the dark current noise may easily outweigh the positive effect of the light guide on the signal-to-noise ratio. The dark current noise may be several times larger than necessary when a light guide is situated too close to the photocathode. With a signal registration system which allows the PM photocathode to be kept at ground potential, e.g., photon counting (cf. Section V.D), the light collection efficiency can be maximized by terminating the light guide close to the photocathode of the PM tube.

Even if one cannot terminate a light guide close to the photocathode, it may be possible to increase the light collection efficiency with a light guide. In one commercial TLD readout unit an increase of 70% was obtained even though the light guide had to be terminated 6 mm from the photocathode.[5] The light guide simply consisted of a perspex® tube with silver tape on its inside.

A lens can be used to concentrate the TL from the readout chamber on the light detector. A light detector with a smaller area than the readout chamber cross section can then be used. This minimizes the dark current from the light detector, since it is proportional to the area of the light detector. However, some light will be lost because the lens extends only over a limited solid angle as seen from the TL material. The best situation in which to use a lens is together with a parabolic readout chamber (cf. Section IV.A). Such a light collection unit can be used when the TL material is lifted out of its sample changer and heated in a jet of hot gas (cf. Section III.C).

V. LIGHT-MEASURING SYSTEMS

A. Introduction

The light-measuring equipment in a TL reader consists of a photoelectric light detector in which light is converted into electric charge, amplifiers for this electric signal, and a signal registration unit that quantifies and stores the electric signal. In PM tubes the light detector and an amplifier are integrated but sometimes an external amplifier is also required. The optimal signal registration method will depend on the TL application. In many TL measurements it is sufficient to measure the total charge from the PM tube during readout, while in other cases it is also of interest to record the variations of the TL signal with time during readout.

A PM tube is the most commonly used light detector in TL readers. There are many pitfalls in the use of PM tubes and it is evident that improper handling in many cases is the cause of poor precision in TL measurements, especially in low dose measurements. For this reason they are given a rather detailed treatment in the following sections.

B. PM Tubes
1. General Principle

The most sensitive device available to detect light is the photomultiplier (PM) tube. A PM tube consists of a photocathode and several (6 to 14) dynodes in an evacuated glass tube. When the photocathode is illuminated photoelectrons are emitted. The photoelectrons are accelerated towards the first dynode with the aid of a potential difference between the photocathode and the first dynode. When they impinge on the dynode, which consists of a metal plate with a secondary emission coating, they lose their energy in the coating and secondary electrons are emitted. These electrons are accelerated towards the second dynode where they in turn liberate secondary electrons. In this way an electron avalanche propagates through the tube and causes a charge pulse to appear at the anode of the PM tube. Almost every electron liberated at the photocathode and reaching the first dynode gives rise to a charge pulse. A few electrons do not because the emission of secondary electrons from a dynode is a stochastic process and the probability of zero emission is finite.

The pulse nature of the output signal from a PM tube makes its measurement possible in two ways: one can either measure the charge or count the number of pulses (photon counting). Both signal registration methods will be treated in some detail in Section V.D.

The stochastic nature of the secondary electron emission from the dynodes causes the output pulses at the PM anode to have a pulse charge distribution even when all pulses are started by single electrons from the photocathode. The shape of the pulse charge distribution is mainly determined by the first few dynodes[68] in the tube. One should therefore not hesitate to use high gain PM tubes with many dynodes when measuring small TL signals because the noise due to the pulse charge distribution is not higher in such a tube than in one with fewer dynodes. Provided the high voltage supply is stable enough it is generally better to amplify the pulses in a high gain PM tube with many dynodes than to use an external amplifier and a PM tube with fewer dynodes.

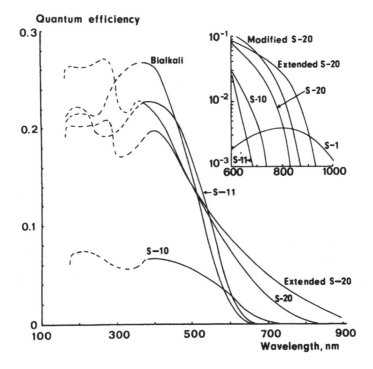

FIGURE 8.   PM tube photocathode quantum efficiencies for different photocathode materials (EMI). S-1 AgOCs, S-10 BiAgOCs, S-11 Cs$_3$SbO, S-20 Na$_3$KSbCs, and Bialkali K$_2$CsSb. Solid curves apply to all photocathode window materials, while broken curves apply only to quartz windows.

2. *Quantum Efficiency*

The photocathode in a PM tube consists of a photo-emissive material coated either on the inside wall of the evacuated glass tube or on a metal plate inside the tube. In TL readers end window photocathodes are most often used. Photocathodes of the end window type are generally semitransparent for light because the photocathode coating has to be thin enough to allow the released photoelectrons to escape.

The PM tube quantum efficiency, η, is defined as the fraction of the photons incident on the PM tube photocathode that give rise to photoelectrons leaving the photocathode. It varies with the photocathode material, the photocathode temperature, and the wavelength of the light. The wavelength dependence is determined by the light transmission properties of the glass in front of the photocathode and by the photocathode material. The transmission properties of the glass mainly affects the quantum efficiency in the blue and ultraviolet wavelength region while the choice of photocathode material mainly affects the red sensitivity of the PM tube (Figure 8).

Because of the wavelength dependence of the quantum efficiency, the photocathode of the PM tube should be chosen so that it "matches" the TL emission spectrum. The quantum efficiency has a maximum at 400 nm for most photocathode materials and it is in the red and infrared wavelength regions that the largest differences between cathode materials are found. The use of red-sensitive PM tubes should be avoided unless the TL emission requires it, otherwise incandescent light may introduce unnecessary noise. Remember that incandescent light with wavelengths above 700 nm may contribute significantly to the measurements (cf. Section IV.B.1). Furthermore, the dark current is generally largest for red-sensitive PM tubes (cf. Section V.C.2). The differential quantum efficiency with respect to the wavelength varies with the temperature in such a manner that both the spectral response and

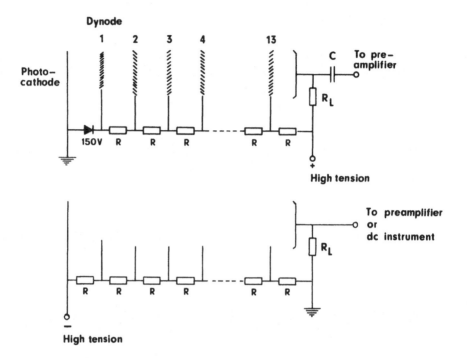

FIGURE 9. Examples of PM tube dynode resistor chains. The top chain can be used only for photon counting, while the lower chain can be used both for photon counting and charge measurements.

the photocathode sensitivity vary.[69-71] Variations of up to 1%/C° have been reported. Cooling the photocathode generally increases the quantum efficiency in the blue wavelength region and decreases it in the red region.

It is known that some photocathodes tend to decrease in sensitivity upon extended exposure to high light levels. The loss in sensitivity, often called photocathode fatigue, can be either temporary or permanent. Therefore, one should definitely not expose a PM tube to room light with the bias voltage on.

3. Collection Efficiency

The fraction of the photoelectrons leaving the photocathode that reaches the first dynode, the collection efficiency, is around 0.9 for PM tubes of the box and grid type, while it is only about 0.6 to 0.7 for PM tubes with dynodes of the venetian blind type. In this respect, box and grid PM tubes are to be preferred. However, the increase in signal-to-noise ratio will be less than the increase in collection efficiency because the light, the dark, and the background signals all increase as the collection efficiencies increase. Other factors may therefore be more important when choosing a PM tube.

4. Dynodes and the Resistor Chain

PM tubes can be constructed with several types of dynodes: venetian blind, box and grid, linear focused, and of circular focus. The main advantage with dynodes of the venetian blind type is that they can be made with large areas so that it is easy to obtain a high collection efficiency from large photocathodes. This is one reason why they are common in PM tubes in TL readers. The rise time and the full width at half maximum (FWHM) for pulses starting as single electrons from the photocathode is, in PM tubes with venetian blind dynodes, of the order of 15 and 30 nsec, respectively. This is faster than for box and grid dynodes but considerably slower than for linear-focused dynodes which have rise times and FWHM of the order of 2 nsec. Linear-focused dynodes are therefore better suited for photon counting techniques because the fast pulses allow a wider dynamic range to be utilized. The fast rise time, however, puts severe constraints on the design of the output circuit and venetian blind dynodes are therefore preferred if high count rate problems are not expected.

The width of the output pulses at the anode is determined by differences in arrival time of the electrons constituting the pulse. The differences appear because the electrons follow different trajectories through the tube.

The secondary emission from a dynode is dependent on the energy of the impinging electron. It increases with increasing electron energy and the mean dynode multiplication factor, g, varies with the potential difference between the dynodes with a power between 0.7 and 1.0 for most dynode materials. The total gain, G, of a PM tube with n dynodes can be written as

$$G = g^n \qquad (2)$$

The gain of a PM tube is therefore determined by the number of dynodes and the potential differences between them. The potential differences between the dynodes are in practice obtained by connecting them to different points in a series of resistors connected to a high voltage power supply in such a way that the potentials of the dynodes increase towards the PM tube anode (Figure 9).

Increasing the potential difference across the dynode chain therefore increases the PM tube gain which may be as high as 10^6 to 10^8. Since the probability for a pulse to fail to propagate through the tube is mainly determined at the first dynode, one often uses a zener diode instead of a resistor between the photocathode and the first dynode. The zener diode keeps the potential difference between the cathode and the first dynode constant irrespective of the value of the potential difference across the whole dynode chain. By correct choice of zener diode the cathode to first dynode potential difference can always be kept at its optimum.

The multiplicative nature of the charge amplification in a PM tube makes the gain, G, very sensitive to changes in the potential difference across the tube. Differentiation of Equation 2 shows that the relative change in the gain may be n times larger than the relative change in the potential difference across the tube, where n is the number of dynodes. A 0.1% change in potential difference might therefore cause a 1% change in PM tube gain. It is therefore important to use stable, high quality, high voltage power supplies for bias of PM tubes. The optimum bias voltage for a PM tube depends on the signal registration system and this subject will therefore be treated together with the signal registration systems (Section V.D).

5. Electric and Magnetic Shields

All PM tubes are sensitive to some extent to electric and magnetic fields. These fields change the electron trajectories through the tube which may result in a change in tube gain. The electrons moving between the photocathode and the first dynode are particularly sensitive to external fields. It is therefore necessary to shield PM tubes from external electric and magnetic fields. Both types of shielding are achieved at the same time if a mu-metal shield at photocathode potential surrounds the PM tube envelope. Since the photocathodes are often biased at a negative potential, it is important to take precautions with regard to the electric shock hazards introduced with such a shield.

The effect of magnetization of the metallic structure in PM tubes caused by the heater current in a TLD reader with ohmic heating has been thoroughly investigated by Saunders.[72] He found variations of up to 9% in the light source readings caused by the final pulse of the preceding alternating heater current. Depending on the direction of this pulse, the light source reading was either increased or decreased. He also found that the PM tube sensitivity

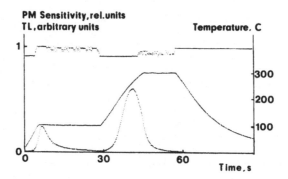

FIGURE 10. Variations of the PM tube sensitivity (top curve) due to an alternating heating current. The sensitivity variations were observed during a heating cycle (middle curve) used for readout of $Li_2B_4O_7$:Mn dosimeters. The lower curve shows the TL as a function of time during readout.

was directly affected while the heater current was flowing. To eliminate the sensitivity variations Saunders had to increase the distance between the heating element and the PM tube to 9 cm.

The problems described by Saunders have also been studied in some detail by others.[73] A reader in which the last pulse of the alternating heating current always was in the same direction was then used. This caused the PM tube sensitivity to decrease between consecutive readouts.[73] Saunders observed both increases and decreases because the direction of the last heating pulse varied.

Figure 10 shows how the PM tube sensitivity varies during readout of a $Li_2B_4O_7$:Mn dosimeter with a planchet heating element of the same type as studied by Saunders. For comparison the heating cycle and the TL glow curve of $Li_2B_4O_7$:Mn are included in Figure 10. During periods when the temperature increased and the heating current was kept on continuously, the PM tube sensitivity decreased by 10%. During periods with constant temperature the heating current was intermittent and the sensitivity varied between 1.0 and 0.9. Also in this case it was possible to get rid of the sensitivity variations by moving the PM tube away from the heating element.

C. PM Tube Dark Signal

1. General Aspects

Several processes can induce anode signals in an unilluminated PM tube. Some of these processes give rise to pulses starting as single electrons, while others give rise to large charge pulses or correlated pulses (afterpulses). Since leakage currents contribute only to the dark current there need not exist an unambiguous dependence between the dark current and the dark pulse rate (cf. Section V.D.5). The presence of correlated pulses makes it impossible to assume *a priori* that the number of dark pulses obey Poisson statistics. The noise in the dark current should therefore be estimated from repeated measurements whenever a PM tube is used in measurements limited by this noise.[74,75]

2. Thermionic and Field Emission

Two processes that initiate dark signals are thermionic and field emission from the photocathode and the dynodes. The emission rate of electrons depends on the work function and the temperature of the material as well as on the electric field strength in the tube. The work functions of red-sensitive photocathodes are lower than for other photocathode materials. The dark signal is therefore generally larger in red-sensitive PM tubes than in other

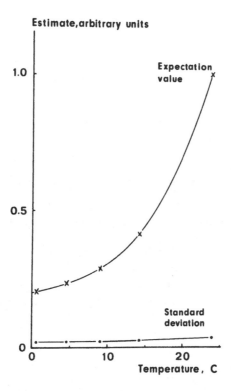

FIGURE 11. Estimated expectation values and standard deviations for the dark current from an EMI 9824 PM tube. The estimates were obtained from repeated 10-sec integrations.

PM tubes. Red-sensitive PM tubes should therefore not be used in TL readers unless the TL emission spectrum is in the red wavelength region.

The dark signal due to thermionic emission can be reduced by cooling the PM tube, thus reducing the noise in the dark current. Figure 11 shows estimated values of the standard deviation and the expectation value for the integrated dark current from an EMI 9824 PM tube as a function of the tube temperature. The estimates were derived from 300 consecutive 10-sec integrations of the dark current at each temperature. The EMI 9824 tube is relatively red sensitive and was used for readout of $Li_2B_4O_7:Mn_1Si$ dosimeters. Figure 11 reveals that the largest advantage of cooling the PM tube might well be that the dark current expectation value becomes considerably less sensitive to variations in the PM temperature. In fact the noise in the dark current (the standard deviation) decreased by only 4% by cooling from 10.0 to 0.5°C.

For some tubes with bi- and trialkali photocathodes the dark signals level off to approximately constant values below -20°C.[69-71] It is believed that thermionic emission is negligible below this temperature and that the residual dark signal is due to other processes. The residual dark signal has a noise which is larger than predicted from Poisson statistics and it is therefore questionable to cool these tubes below 0°C.

When a PM tube is operated below room temperature, it should be kept in mind that the quantum efficiency of the photocathode depends on the temperature.[69-71] Under certain conditions an uncooled tube can therefore give a better signal-to-noise ratio than a cooled tube.[80]

A disadvantage with PM tube cooling is that condensation of water vapor may cause problems. Condensation can, however, be avoided by proper isolation and air-tight housing of the PM tube.

In some TL readers the PM tube housing is cooled. In such readers it is important to realize that the PM tube temperature may be considerably higher than that of the housing. The temperatures may differ by as much as 10°C. This means that optimal thermal contact between the PM tube and the cooling element may increase the temperature of the PM tube housing by 10°C without increasing the dark signal noise and may reduce problems with condensation on the PM tube housing. However, a cooling mantle cannot be put in direct contact with the PM tube near its photocathode without increasing the dark signal unless the photocathode is kept at earth potential (cf. Section V.C.4).

It is known that exposure of the photocathode to high light levels increases the dark signal. The increase is observed even if the high voltage is not applied to the tube during the illumination. If the high voltage is applied to the tube shortly after such an illumination, the dark signal is observed to decrease with time and it may return to its equilibrium value after several hours. The increase in the dark current is probably caused by thermal emission of electrons trapped at metastable energy levels during the illumination. Because of this effect it is necessary to allow the dark current to stabilize whenever a PM tube has been exposed to light. When PM cooling is utilized the stabilization will proceed faster if the cooling is turned off during the stabilization.

The dark current increase is larger after exposure to blue than to red light. Handling of PM tubes after fluorescent lighting should therefore be avoided and subdued incandescent lamps are preferred.

3. Interactions of Ionizing Radiation

Ionizing radiation, e.g., β particles from disintegrations of ^{40}K in the glass envelope, cosmic rays, and γ-radiation from radioactive nuclides in the environment can produce dark signals, either directly by interactions in the photocathode and the dynodes or indirectly by Čerenkov light, scintillations, and phosphorescence in the glass envelope of the PM tube.[81-85] The emitted light is in general a more important source of dark pulses than direct interactions in the cathode and the dynodes.[81] The dark pulse rate from the above processes depends on the positioning of the photocathode and its area.

Čerenkov light emission takes place in such a short time interval that the different photons emitted during the slowing down of one charged particle are not resolved. Čerenkov light may therefore produce pulses which on the average are larger than those due to single photons. Čerenkov light emission is therefore most severe in charge measurements (cf. Section V.D).

Scintillations and phosphorescence, on the other hand, last long enough to allow the resolution of different light photons caused by a single energy deposition event when photon counting techniques are used. The resolving time of the measuring equipment therefore often determines if, after each interaction, a number of correlated pulses with a pulse charge distribution equivalent to that of single photons, or one pulse with a larger charge or, more probably, something between these two extremes is observed. The noise in the dark signal may be several times larger than that predicted from Poisson statistics because of the cosmic-ray afterpulses.[86]

Interactions of ionizing radiation in other glass parts of the optical system may also induce significant background signals. Thick solid light guides should therefore be avoided. For example, in a quartz light guide with an area of 15 cm² cosmic-ray muons at sea level cause about 15 Čerenkov light pulses per minute.[87] These pulses may cause considerable noise in charge measurements because they are much larger than the single TL photon pulses, especially when the light guide is several millimeters thick. It has been calculated that a Čerenkov light pulse produced by a relativistic charged particle in a 10-mm thick glass plate in front of an EMI 9623 PM tube on the average produces 80 photoelectrons at the photocathode.[87] The EMI 9623 PM tube has an S-11 (Cs_3SbO) photocathode. The above figure

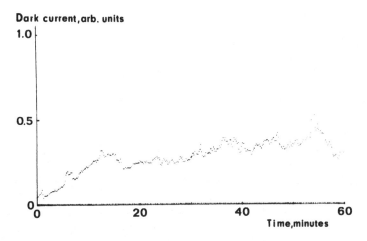

FIGURE 12. The dark current as a function of time after application of the potential difference across the dynode resistor chain of an EMI 9824 PM tube biased at −1200 V. Top curve: plastic support not at cathode potential touching the rim of the PM photocathode end of the tube. Bottom curve: without the plastic support.

is valid when the relativistic particle travels towards the photocathode. In order to minimize the Čerenkov pulses, the PM tube should not be sensitive to UV light unless the TL emission takes place in the UV region. Another method that may also reduce other types of radiation-induced light emission from glass parts is to place optical filters between the light guide and the PM tube instead of in front of the light guide.

Apart from environmental radiation, man-made radiation sources may also introduce dark signals, e.g., β-radiation from reference light sources containing $^{90}Sr/^{90}Y$ used in some TL readout equipment may induce luminescence in the glass parts of the optical system.[17]

4. Photocathode Disturbance

The dark signal is minimized if the PM tube is surrounded with a mu-metal shield at cathode potential. The dark signal and its noise can reach values several orders of magnitude larger than necessary if the shield potential differs from the cathode potential. Also the light-induced signal becomes noisier. Although this information is well known, TL instruments are still delivered with inadequate electrical shielding. The mechanism responsible for the increase in the dark signal is not known in detail. Electroluminescence has been suggested as one possible source. However, it is not sufficient to keep the PM tube shield at cathode potential to avoid increasing the dark signal noise. It is also necessary to insure that all objects not at cathode potential are sufficiently far away from the photocathode. Any object, e.g., a light guide, in contact with the photocathode increases the dark signal tremendously if not at cathode potential. This happens even when the object is an insulator. If an object not at cathode potential touches the photocathode end of a PM tube for a long time it may permanently damage it.

Not only objects touching the photocathode end of a PM tube, but also objects close to the photocathode may increase the dark signal if their potentials differ from that of the cathode. A distance not less than 5 mm is recommended between the photocathode and any object not at cathode potential.

The problems that may occur due to cathode disturbance cannot be overemphasized. Figure 12 shows how the dark signal integrated during 10-sec intervals varied with time in a TLD reader where the outer rim of the PM tube rested on a plastic support. The photocathode was biased at -1200 V relative to ground. Included in Figure 12 is the dark signal with the plastic support removed. It can be seen that the dark signal noise with the support is several hundred times larger than without support.

One method to avoid cathode disturbance is to bias the PM tube with the photocathode at ground potential so that light guides in contact with the PM tube can be used. However, one then either has to perform floating charge measurements at anode potential or rely on photon counting techniques (cf. Section V.D).

5. Ion Pulses

Gas atoms in PM tubes may be ionized by electrons in a charge pulse propagating through the tube. Such ions can form a detectable pulse if they reach the photocathode or a dynode.[86] Ion pulses appear as after pulses and they often have a larger charge than single photon pulses. One source of afterpulses is helium atoms which can diffuse into the PM tube. Coates,[86] e.g., found that the main source of afterpulses in tubes more than 2 years old was helium. Very old PM tubes should, therefore, be avoided in measurements of small light levels, at least if charge measurements are performed. Furthermore, He should be avoided for purging the readout chamber unless the PM tube housing is sealed since it can penetrate the tube and spoil the vacuum.

6. Leakage Currents

Leakage currents may cause a serious problem if the PM tube is cooled below room temperature and condensation of water vapor occurs. However, leakage currents are no problem at moderate temperatures if modern PM tubes and teflon® sockets are used. To minimize the risk of leakage currents due to condensation of water vapor, it is advisable to use a sealed PM tube housing when PM tube cooling is utilized. This is especially important in humid climates.

7. Other Effects

Fast potential transients may, in many cases, increase the dark signal temporarily. The time required to stabilize the dark signal is therefore minimized if the potential difference across the dynode resistor chain is increased continuously to its operational value instead of applied instantaneously. If a continuous increase is impossible, an increase in many small steps is preferred over one large step. In this connection it should be pointed out that both the PM tube sensitivity and the dark current are best stabilized if the high voltage across the dynode resistor chain is kept on all the time instead of being turned off between different measurement series.

D. Signal Registration
1. General Aspects

The purpose of the signal registration system is to quantify and store the signal from the light detector. PM tubes are by far the most common light detector in TL readers and apart from some general remarks and a subsection on glow curves (Section V.D.2), this section on signal registration will be devoted to PM signal registration systems. Some of the principles outlined here are, however, also applicable to other photoelectric detector systems, especially those concerned with charge measurements.

The PM tube signal consists of charge pulses initiated by single electrons. This discrete nature of the PM signal makes it possible to measure it by one of two main methods: the charge can be measured or the number of pulses counted. The latter method is generally denoted photon counting. It has been claimed that photon counting should result in signal-to-noise ratios several times or even orders of magnitude larger than charge measurements.[89-94] Spanne[5] has proved that this is not the case when both techniques are applied in optimal fashion. Both signal registration methods are treated in some detail below. In both methods the PM signal can be measured either by integrating during the whole or a selected part of the readout cycle or by recording it as a function of temperature or time during the readout. The term "glow curve" is usually used to denote the TL as a function of temperature during readout.

2. *Glow Curves*

For many routine TL dosimetry measurements it is sufficient to integrate the TL signal during a selected part or the whole of the readout. For research purposes it is often of value to record glow curves. In practice the TL is often recorded as a function of time instead of temperature and the corresponding curve is also denoted as a glow curve because of difficulties to determine the temperature of the TL material during readout. Becker[95] has discouraged the publication of such curves in research papers, claiming it to be a confusing habit. Certainly it is best to present the TL as a function of temperature in many cases. However, in some cases it is justified to present the TL as a function of time instead of temperature, e.g., when a heating cycle with temperature plateaux is used. If possible, the heating cycle should then be included in the figure.

Apart from being dependent on the TL material the glow curve shape depends on several other parameters, some of which may be related to the readout equipment. The glow curve may be affected by the readout heating rate, the previous thermal treatment inclusive of the cooling rate of the TL material, the properties of the radiation-material interaction, the absorbed dose, earlier accumulated absorbed dose and nonradiation-induced signals from the TL material, and the readout equipment. All these parameters affect the result independent of whether the PM signal is recorded as a function of temperature or time during readout. The heating rate dependence for some materials manifests itself during both recording modes because the glow peak temperature shifts to higher temperatures with increased heating rate, while the glow peak heights decrease due to thermal quenching. The heating rate should therefore be specified when glow peak temperatures or glow curves are presented.

Research on TL materials usually requires glow curve recording. The readout heating rate should then be linear and of the order of 1°C/sec to minimize the thermal gradient and optimize glow peak resolution. At higher heating rates the glow peaks may overlap due to thermal lag. Linear heating rates also greatly simplify the temperature-time relationship in readout so that direct readout of the glow curve is facilitated. Glow curve recording is also of benefit in routine dosimetry measurements when signal integration techniques are used to quantify the TL emission. For example, it is possible to detect instrumental malfunctions such as changing thermal contact between the heating medium and the dosimeter or anomalous appearance of nonradiation-induced signals.

When signal integration is utilized it is often advantageous to limit the integration to a selected part of the readout cycle. A lower integration limit can exclude low temperature TL peaks that contribute to the fading. Optimal separation of the low and high temperature peaks can be achieved using a low temperature plateau before starting the signal integration. An upper integration limit is of value because nonradiation-induced signals from the dosimeter and the readout equipment can then, at least to some extent, be excluded from the measurements. A lower integration limit also may reduce nonradiation-induced signals, e.g., the spurious signal appearing very early during readout with a jet of hot nitrogen (cf. Section III.C).

The ultimate lower limit on the signal integration interval is when the so-called peak height method is used. With this signal quantification method the PM signal is recorded as a function of time and the height of the glow peak measured. This means that an integration interval related to the time constant of the recording equipment is used. In principle, the merits of the peak height method should be the same as those which favor the introduction of a limited signal integration interval. Whether or not one uses the peak height method is therefore a question of optimizing the length of the integration interval.

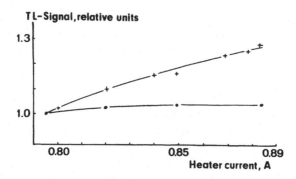

FIGURE 13.   The TL signal from a LiF:Mg,Ti dosimeter as a function of the heating current through a metal planchet. (·) Integrated area under the glow curve. (+) Peak height. The measured values have been normalized to the values at 0.795 A.

It is unquestionable that the peak height method reduces the influence of the fading of low temperature TL peaks. The other advantage claimed to apply to the peak height method is that an increased signal-to-noise ratio should improve the signal-to-noise ratio in low dose TL measurements. This is far from evident (cf. Equation 1) and true only under certain circumstances. First of all, the glow curve for the nonradiation-induced signals must differ from that of the TL. This is the case in many TL measurements because nonradiation-induced signals often appear as a high temperature tail. Furthermore, the noise in the measured nonradiation-induced signals must not be dominated by quantum noise. If these two requirements are fulfilled, the peak height method may result in a higher signal-to-noise ratio in low dose measurements than any other signal integration interval even though only a small fraction of the TL is used with the peak height method. However, in such a situation it may be possible to reduce the noise in the nonradiation-induced signals by other means.

Another situation where the use of the peak height method may be valuable is when the nonradiation-induced signal varies in a systematic although unpredictable way between different readouts. The effects of such variations are minimized when the signal-to-background ratio is maximized.

Whenever the peak height method is used and the PM signal is recorded as a function of time during readout, it is important to reproduce the heating cycle. Any heating cycle variation causing a change in the rate of heating will affect the peak height. With a linear heating rate, a 1% change in heating rate causes a 1% change in peak height. The heating rate dependence on the peak height is one of the main drawbacks of the peak height method as compared to the integration method, which can be made essentially independent of small changes in the heating rate (Figure 13). The other main drawback limiting its useability is that only a small fraction of the TL is used.

3. Charge-Measuring Techniques

In the past most TLD readers were equipped with signal registration systems measuring the output charge from the PM tube. In most cases, however, both pulse and charge integration measurements are possible. The popularity of the charge-measuring technique is due to the fact that it is a simple, well-known technique and that charge-measuring systems can be constructed with instruments available in most dosimetry laboratories.

The principle for charge measurements is simple: the PM anode, biased at ground potential, is directly connected to a current-measuring or charge-integrating instrument. This means that standard DC techniques can be used. Since the PM current may be less than 10^{-9} A, electrometers are often used for these measurements. The main disadvantage with the charge-

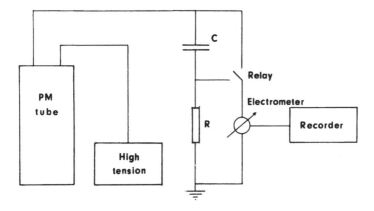

FIGURE 14. An example of instrumentation for charge measurements using DC techniques. Both charge integration and glow curve recording are possible.

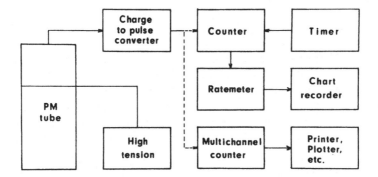

FIGURE 15. Schematic diagram of a charge-measuring system using a charge-to-pulse converter. A multichannel analyzer may replace the counter and the ratemeter.

measuring technique is that it is necessary to bias the PM anode at ground potential unless the charge measurement can be made with the instrument floating at a potential as high as +1000 V.

Figure 14 shows the simple principle for a signal registration system that allows both glow curve recording and charge integration. During readout the relay connects the PM anode to the capacitor which is charged by the PM current. At the end of the signal integration the relay is switched back and the voltage is measured with an electrometer which during the integration measures the potential difference across the resistor caused by the current charging the capacitor. After each measurement the capacitor is discharged by short circuiting the input of the electrometer across the capacitor.

In Figure 15 another charge-measuring system is schematically presented. In this set-up the PM current is integrated and converted to charge pulses with the aid of a charge-to-pulse converter. The charge-to-pulse converter integrates the PM current and each time the integrated charge reaches a certain fixed level a standardized pulse appears at the output. The charge integrator is then reset and the integration started again. In this way the PM current is converted to a pulse train. The signal can then be integrated by counting the number of pulses or the glow curves recorded with a ratemeter. An alternative way to record glow curves is to use a multichannel pulse height analyzer operated in a multiscale counting mode. This recording method has the advantage that it yields digitized glow curves fit for

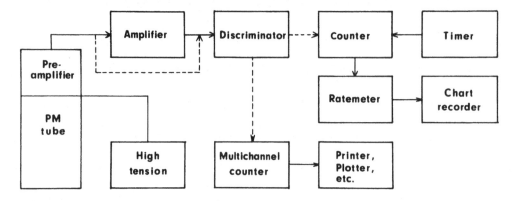

FIGURE 16.   Schematic diagram of a photon counting system. A multichannel analyzer may replace the counter and the ratemeter.

computer analysis. Since the time interval between the output pulses from the charge-to-pulse converter depends on the input current, they are often called current-to-frequency converters, a slightly misleading name.

Charge-to-pulse converters are often equipped with circuits backing off their inherent leakage currents. These can also be used to back off the PM dark current by balancing the backing circuit so that the mean value of the dark signal after the converter becomes zero. With such backing it is important to realize that all signal values less than zero are put equal to zero so that experimental estimates of the dark current noise may indicate a smaller influence of the dark current noise on the signal-to-noise ratio than is actually the case when low light levels are measured. Backing off the dark current in no way decreases the noise in TL measurements.

DC techniques are used in most charge-measuring systems in TL readers. However, Lippert and Mejdahl[96] have reported using an AC signal from the PM tube by modulating the light with a chopper. The chopper consisted of a rotating disk with cutouts inserted in the light path between the readout chamber and the PM tube. The chopping frequency was approximately 70 c/sec. The resulting AC signal was amplified and detected by a phase-sensitive detector. Lippert and Mejdahl[96] found that a modulation of the light signal effectively suppressed the noise caused by the dark current in the PM tube.

When trying to evaluate the noise reduction of this chopping technique two things are important to consider. The first is that half of the light signal is lost, and the other is that it is mainly the low frequency noise and the effects of systematic changes in the dark current that are reduced. Noise with higher frequencies than the chopping frequency are not reduced at all. The two effects counteract on the signal-to-noise ratio and it therefore seems worthwhile to use chopping only when one suspects that the dark current changes systematically, e.g., because of heating of the photocathode. This is in fact the main cause for the decrease in the dark current noise observed by Lippert and Mejdahl.

4. Photon Counting Techniques

In the photon counting technique, the PM anode charge pulses started by single photons interacting in the photocathode are counted. Figure 16 shows a schematic set-up for photon counting. When photon counting is used it is possible to bias the photocathode either at ground or at negative potential. With the photocathode at ground potential it is necessary to decouple the anode with a high voltage capacitor since it is then at a high positive potential. Figure 9 shows a resistor chain suitable for use in photon counting with a PM tube of the venetian blind type. The design of the output part of the dynode resistor chain is extremely critical, especially the choice of anode load resistor. The faster the output pulses the more

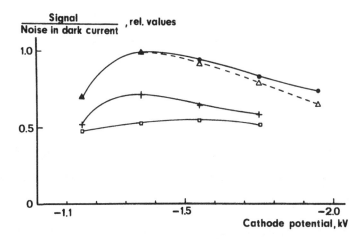

FIGURE 17. The ratio of light signal to dark current noise for the EMI 6097S PM tube. Photon counting: (▲) Theoretical with minimum discriminator. (·) Theoretical with optimum discriminator. (+) Experimental with optimum discriminator. Charge integration: (□) Experimental.

critical is the output circuit which must be designed to minimize ringing and reflections. The output circuit of the PM tube is followed by a preamplifier which reduces the influence of stray capacitance and pick-up in the cables leading to the amplifier or counter. All preamplifier connecting cables should be as short as possible and the preamplifier rise and fall times must be comparable to the PM output pulse shape, i.e., a few nanoseconds. A fast amplifier is occasionally used after the preamplifier, but it is usually preferred to use the PM tube amplification (cf. Section V.B.1). After the amplification stage a fast discriminator must be used to reject small charge pulses in the dark current, otherwise the photon counting method may become inferior to charge measurements.[5] An upper level discriminator rejecting large pulses, which almost exclusively are dark pulses in new PM tubes, may also be used. Horowitz[97] has suggested that the signal-to-noise ratio in nonoptimized photon counting systems (e.g., with large solid light guides) may be significantly increased by using an upper level discriminator.

In the optimized measurements reported by Spanne,[5] however, the signal-to-noise ratio would have increased by only a few percent with the introduction of such a discriminator since large dark pulses made only a small contribution to the total number of dark pulses. The largest gain in signal-to-noise ratio due to reduction of noise caused by large pulses is obtained when changing from charge integration to photon counting.

When the photon counting technique is used, glow curve recording has to be done with a ratemeter or a multichannel counter, e.g., a pulse height analyzer operated in the multiscale counting mode.

5. Optimizing the Potential Difference Across the PM Tube

The limiting factor governing the ability of a properly operated PM tube to detect low light levels is the dark current noise. It is therefore appropriate to use the light signal-to-dark current noise ratio as a criterion when optimizing the potential difference across the dynode chain. As this ratio is a complicated function of many variables, it is in practice impossible to determine theoretically from a mean value of the dark current and should be determined from experimental estimates of the expectation value of the light signal and the dark current noise. The estimates are obtained from repeated measurements by standard statistical techniques. As an estimate of the dark current standard deviation, the sample standard deviation, σ_{DP}^*, should be used and defined as

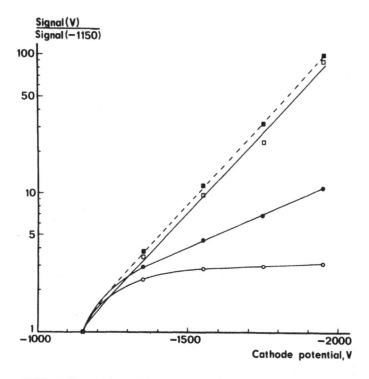

FIGURE 18.   Light and dark signals normalized at -1150 V as functions of the cathode potential relative to ground for an EMI 6097S PM tube. Open symbols = light signal; closed symbols = dark signal. Squares = charge integration; circles = photon counting.

$$\sigma_{DP}^{*} = \left(\frac{1}{n-1} \sum_{i=1}^{n} (x_i - \bar{x})^2 \right)^{1/2} \tag{3}$$

where n is the number of measurements; x_i the measured values; and $\bar{x}$ the mean value of the sample.

Observe that the ratio of the number of light-to-dark pulses is not an appropriate criterion for describing the ability of a PM tube to detect low light levels. Instead Equation 1 should be used.

Figure 17 shows estimated values of the ratio of light signal-to-dark signal noise for an EMI 6097S PM tube as a function of the cathode potential. A linear dynode chain was used in the measurements. It can be seen that the ratio varies with the cathode potential for both charge integration and photon counting and that the optimum cathode potential is different for the two signal registration methods. The signal-to-noise ratio for photon counting first increases with decreasing cathode potential. This is caused by an increase in the number of pulses detected with increasing gain in the PM tube. However, the number of dark pulses detected increases faster with the gain than the number of signal pulses (Figure 18), and the signal-to-noise ratio thus reaches a maximum and then decreases with decreasing cathode potential. The charge integration signal-to-noise ratio changes more slowly with the cathode potential. Figure 17 shows that the signal-to-noise ratio is larger for the photon counting method than for the charge integration method. However, the difference is not larger than 40% for any cathode potential.

Included in Figure 17 are theoretical estimates of the signal-to-noise ratio for photon counting. These were calculated assuming that the pulse numbers followed a Poisson distribution and used the measured light and dark signal values employed to derive the curves

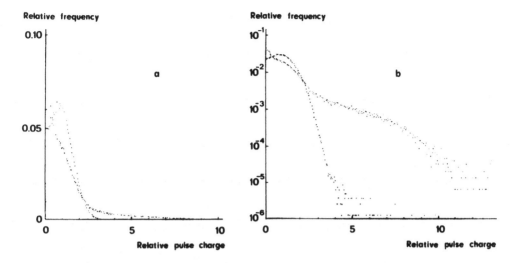

FIGURE 19. Light pulse (peaked curve) and dark pulse charge distributions for an EMI 97890 PM tube. (a) Linear and (b) logarithmic ordinates.

for photon counting in Figure 18. Theoretical values were derived both for optimum and minimum discriminator levels. It is evident that the experimental values for the signal-to-noise ratio are lower than the theoretical. This is caused by correlated pulses not considered in the theoretical curves and shows the importance of using experimental estimates rather than theoretical values when specifying the performance of the light-measuring equipment.

Figure 17 and other measurements[98] have shown that the optimum cathode potential for two PM tubes of venetian blind type was less than − 1000 V when the charge integration technique was used. However, in measurements of small absorbed doses, TLD readers are usually operated with cathode potentials higher than − 1000 V. This may be due to inappropriate construction of the PM tube housing. Objects not at cathode potential too close to the photocathode may increase the dark current noise drastically. The magnitude of the increase depends on the potential difference between the object and the photocathode which may shift the optimum cathode potential to higher values than for properly operated PM tubes.

With respect to systematic uncertainties, Figure 18 shows that for charge integration the same relative change of the cathode potential causes the same relative gain change independent of the cathode potential. For photon counting, on the other hand, the light signal changes relatively slowly with the cathode potential at high potential. The dark signal for photon counting varies more than the light signal, but is still less dependent on the cathode potential than during charge integration.

The light signal pulse rate increases more slowly since the light signal pulse charge distribution contains relatively less small pulses than the dark pulse charge distribution (Figure 19). From Figure 18 it can therefore be concluded that the photon counting method is less sensitive to changes in cathode potential than the charge integration method.

6. Comparison of Photon Counting and Charge Integration

Photon counting methods applied to TL measurements have been reported by several investigators,[89-94] all concluding that the photon counting method is superior to the DC method in TL measurements of small absorbed doses. These investigators claimed that the signal-to-noise ratio for photon counting was several times or even an order of magnitude larger than for charge measurements. Spanne[5] has shown that the difference generally is much smaller if the PM tube is operated optimally.

Two main reasons that photon counting should result in a much better signal-to-noise ratio than charge measurements have often been suggested: the noise due to the stochastic nature of the dynode multiplication process is eliminated because all pulses are counted equally irrespective of their charge, and small dark pulses can be excluded form the measurements with a lower level discriminator. The noise in the multiplication process leads to a random charge distribution in which the dark pulse charge distribution contains relatively more small and large pulses than the light pulse charge distribution (Figure 19). This difference is used to increase the signal-to-noise ratio in photon counting measurements by introducing lower and upper level discriminators. However, the increase caused by the latter is small because of the relatively small fraction of large pulses.

The relative noise in the integrated output charge, Q, from a PM tube can be written as

$$\frac{\sigma(Q)}{E(Q)} = \left[\frac{V(N)}{E^2(N)} + \frac{1}{E(N)} \frac{V(q)}{E^2(q)} \right]^{1/2} \tag{4}$$

where N is the number of charge pulses appearing on the anode during a measurement of time t; q is the charge in a pulse ($\ddagger 0$); E(X) is the expectation value of the stochastic variable X; V(X) is the variance of the stochastic variable X; and $\sigma(X)$ is the standard deviation of the stochastic variable X.

Q, q, and N are stochastic variables. Equation 4 shows that the noise in charge integration measurements consists of two parts: one due to fluctuations in the number of detected pulses, the other due to the pulse charge distribution, i.e., due to the dynode multiplication. Equation 4 is valid not only for PM charge measurements, but also for other measurements where a random number of events of random size are summed. A proof of the relation is given by Kellerer.[99] In Equation 4 the effects of the measuring equipment after the PM tube have not been taken into consideration, except through the condition that the measuring time, t, should be much longer than the pulse width. Remember, however, that all current measurements can be considered as charge integration measurements with the time constant related to the integration time.

Using existing models for dynode multiplication, Spanne[5] has shown that the noise due to dynode multiplication decreases the signal-to-noise ratio in charge measurements by at most 40% compared to photon counting when $V(q) = 0$. In many cases the noise increases by less than 40%, the exact figure being determined by the nature of the single photon pulse charge distribution. Furthermore, the exclusion of small dark pulses originating from the dynodes by using a lower level discriminator in photon counting does not result in a greatly increased signal-to-noise ratio. This is so because in charge integration the noise contribution from a pulse is proportional to the square of its charge. Small pulses therefore are automatically discriminated against in the charge integration method. For this reason a lower level discriminator is mandatory in photon counting measurements, otherwise the signal-to-noise ratio may become smaller for photon counting than for charge integration.

Contrary to small pulses, large pulses contribute more to the noise in charge measurements than in photon counting measurements. However, unless unusually large numbers of large pulses are present, the signal-to-noise ratios for the two signal registration methods will not differ by more than a factor of two.[5] Large and correlated pulses are avoided by not using thick light guides, in which ionizing radiation can cause luminescence and Čerenkov radiation (cf. Section V.C.3). Large pulses may also be excluded from photon counting measurements with the aid of an upper level discriminator. However, as indicated in Section V.D.4, the increase in the signal-to-noise ratio will generally be only a few percent unless an unusually large number of large pulses is present.

In conclusion to this rather detailed comparison of PM signal registration methods it can be stated[5] that existing theories describing the noise in the PM output charge indicate that in TL measurements of small absorbed dose the precision in photon counting is better than in charge measurement. On the other hand, it is unlikely that the signal-to-noise ratio for photon counting will be more than twice that for charge integration. In many cases they may differ by considerably less. These findings were experimentally substantiated by estimating the noise from repeated measurements and from measured pulse charge distributions.[5]

The large difference between the precision in photon counting and charge measurements, observed by several investigators,[89-94] can probably be explained by experimental artifacts.[5] If a large difference is observed between the two signal registration methods it is likely that an unnecessary noise source is present and its elimination will not only increase the signal-to-noise ratio for charge measurements, but also for photon counting although the increase will be smaller for the latter method. Clearly, however, from a practical rather than a theoretical point of view, the photon counting technique is preferred due to its greater intrinsic ability to discriminate against unrecognized noise sources.

All the above conclusions are valid when fluctuations in the potential difference across the PM tube contribute negligibly to the noise. From Figure 18 it can be deduced that the photon counting method is less sensitive to such fluctuations. In practice, however, potential difference fluctuations should contribute negligibly to the precision in both signal registration methods if a stable high voltage power supply is used.

It is sometimes claimed that photon counting is inferior to charge integration because the dead time of the counting equipment limits the dynamic range in the former. However, with the fast counting equipment commercially available today, it is the PM pulse width which determines the dead time. For PM tubes with venetian blind dynodes (pulse FWHM = 25 nsec) pulse rates up to 2 MHz can be measured with dead time losses of only a few percent. With linearly focused PM tubes (pulse FWHM = 3 nsec), the maximum pulse rate for dead time losses of only a few percent is about 20 MHz. This results in a dynamic range which is larger than or comparable to that of any existing current integrator. In principle the linear range in photon counting measurements can be expanded by dead time loss corrections, however, these may be of considerable complexity because of the variation of the counting rate during readout. In any event neutral density optical filters are available which can effectively decrease the PM tube counting rate by one to several orders of magnitude.

One of the important advantages of the photon counting method in low dose TL measurements is that it is possible to keep the cathode at ground potential and then to use a light guide in close contact with the PM tube without affecting the dark current noise. This results in good light collection efficiency. It is possible to make charge measurements with the charge measuring instrument floating at more than 1000 V, but setting up such a system may be as difficult as incorporating a photon counting signal registration system.

Another merit of the photon counting method is that it takes advantage of the inherent discrete nature of the PM signal so that further signal processing is unnecessary before transfer to a computer.

E. Other Photodetectors

Good alternatives to PM tubes as light detectors in TL readers are rare. The only realistic alternative seems to be channeltrons with photocathodes. A channeltron is a continuous electron multiplier operating on the same principle as PM tubes but with a different design. It consists of a narrow curved evacuated glass tube whose inner surface is coated with a secondary electron emissive material. By applying a potential gradient along the tube, electrons released from the photocathode are accelerated towards the anode. As the tube is curved the accelerated electrons impinge on the coating of the walls and thereby liberate secondary electrons so that an electron avalanche propagates through the tube. Each photoelectron therefore results in a charge pulse at the anode.

The most obvious advantage of channeltrons as compared to PM tubes is their small size, however, at the present time no comparative studies of stability and precision in TL measurements have appeared in the literature.

Another photodetector is the photodiode. Nonamplifying photodiodes are too insensitive to be a realistic alternative to PM tubes in TL readers.[100] In some avalanche photodiodes it is possible to achieve a charge amplification which makes single light photon detection possible.[101] However, the small sensitive areas of these devices prohibit their use in TL readers. The only situation where photodiodes can compete with PM tubes is therefore in measurements of very high light levels.

F. Reference Light Sources

Many factors affect the light collection and the overall sensitivity of the light-measuring system in a TL reader. Reference light sources are therefore often included in TLD readers to monitor changes in the overall light sensitivity. The reference light sources may either be internal, i.e., built into the reader, or external. In some readers the light source moves into position below the PM tube every time the dosimeter is changed. The light source can then be used together with a feedback system to automatically regulate the high voltage across the PM tube.[11]

The light emission from reference light sources varies with several parameters which must be carefully controlled in order to avoid a situation where corrections in the TL reader sensitivity are incorrectly based on variations in the light emission from the reference light source. Reference light sources often consist of a luminescent material, i.e., a scintillator, and a radioactive material to induce luminescence. As scintillators, both NaI(Tl) and plastic materials have been used. Because long term radiation damage may deteriorate the light emission properties, α emitters should be avoided in reference light sources. Often low energy β emitters like ^{3}H and ^{14}C are used because they cause relatively little radiation damage, the β particles cannot escape the light source, and the long half-lives make decay corrections unnecessary. Because of the low penetration ability of ^{3}H and ^{14}C β particles, the radioactive material has to be mixed with the scintillator. High energy β emitters like ^{90}Y-^{90}Sr and γ emitters should be avoided because the high energy particles may induce luminescence in the glass parts of the optical system of the TL reader.[16]

A light source in which Čerenkov light was generated in a medium by β particles from a ^{90}Y-^{90}Sr source has been described.[51] A shutter was used between the radioactive source and the Čerenkov medium to avoid radiation damage changing the light transmission properties. This effect with the additional risk of inducing luminescence in the glass parts of the TL reader makes such light sources less attractive. The main advantage with the Čerenkov type of light source is that the light emission is independent of the temperature.[51]

The temperature dependence of scintillator light emission must be taken into consideration. Temperature coefficients from -0.05 to -1%/°C have been reported[51,102] for temperatures around 20°C. The lower figure, -0.05%/°C, was observed for a ^{14}C-loaded plastic scintillator. This light source was also found to be sensitive to storage in nitrogen; overnight storage increased the light emission by almost 3%.[102] Plastic scintillators in light sources should therefore be sealed to avoid the affects of the surrounding medium.

When external light sources are used, precautions must be taken to avoid adding phosphorescence induced by room lighting to the light emission. Internal light sources are preferred in this respect. Since the overall light sensitivity depends on the PM tube sensitivity and the light collection efficiency, an ideal reference light source should have the same geometry and transparency as the dosimeters. This is impossible to achieve in practice. Furthermore, the emission spectrum of the light source should resemble that of the TL material because changes in light collection efficiency caused by, e.g., the accumulation of dirt on the readout chamber walls, may vary with the wavelength of the light. Since ideal reference light sources fulfilling the above requirements do not exist, it is preferable to use calibration dosimeters which, apart from a standardized irradiation to a known constant absorbed dose, have been treated identically as the other dosimeters to be read out.

By reading out the calibration dosimeters at regular intervals during the readout of a series of dosimeters, changes in the overall sensitivity of the system can be assessed. An advantage with this method is that the sensitivity of the whole TLD system inclusive of the dosimeters is determined. A new type of reference light source based on a light-emitting diode has been used successfully in a gain stabilization system for scintillation detectors.[103,104] Such a system, however, electronically stabilizes the light emission from the diode and is therefore more complicated than the traditional light sources used in TL readers.

The application of a light-emitting diode in a TL reader has also been described.[105,106] The maximum variation in the light emission during 1 month was found to be about 0.8% when it was stabilized by controlling the potential difference across the diode. However, the reports on the gain stabilization systems for scintillation detectors mentioned above show that the light emission from a diode can be even better stabilized[103,104] so that an overall stability of 0.1% is possible to achieve.

G. Instruments for Measurements of TL Spectra

Measurements of TL spectra are of value since the emission spectrum may give information about the recombination process. Knowledge of the emission spectrum is also required in order to optimize the light detector and the optical filter.

There are two main problems connected with emission spectrum measurements: (1) the low intensity light level behind the monochromator requires heavily irradiated TL samples and (2) the light signal varies with time. The low light levels behind the monochromator are partly due to the differential nature of the measurements. Often, a compromise is necessary between wavelength resolution and sensitivity. The sensitivity problem is considerably reduced when the spectra are integrated over the entire readout cycle, however, the emission spectrum may change with the readout temperature and if this is the case it is necessary to use a spectrometer with a rapidly scanning monochromator.

Three types of rapidly scanning monochromators have been applied to TL measurements. Harris and Jackson[107,108] used a highly sensitive grating spectrometer which allowed emission spectra measurements at low dose in the linear portion of the TL dose-response curve. This is an important achievement in the testing of models relating to supralinearity, e.g., in determining whether the creation of luminescence recombination centers (new optical absorption bands) is contributing to the TL dose response. The Harris-Jackson monochromator consisted of a prism and a rotating mirror. With this type of monochromator the design of the wavelength control system is critical because of the small dispersion of the prism for long wavelengths. Spectrometers with rapidly scanning grating monochromators have also been used,[109-112] but these spectrometers are usually less light sensitive than prism spectrometers. A third type of rapidly scanning spectrometer has been described by Bailiff et al.[113] The monochromator, which consisted of 16 interference filters mounted on the periphery of an aluminum disc rotated at 8 r/sec, had a relatively low wavelength resolution of approximately 10 to 20 nm determined by the bandwidth of the filters. An advantage with this type of spectrometer is that large-area TL samples can be employed.

None of the above instruments are sensitive enough to measure emission spectra for nonradiation-induced signals and therefore, in these measurements, considerably cruder methods (e.g., optical filters[59-61]) must be employed. Since it is often difficult to reproduce measurements of nonradiation-induced TL, two PM tubes must be used simultaneously, one measuring filtered TL and the other measuring unfiltered TL for reference purposes.

An instrument which allows the measurement of emission spectra at a constant temperature without a decrease in TL with time has been described.[114] The instrument requires a large sample of TL material enclosed in a glass tube which slowly passes the entrance slit of a monochromator. The TL material is heated by a resistive heating element in the form of a nichrome® coil around the glass tube. The TL powder is moved along the glass tube by vibrating the tube.

A spectrometer with photographic recording has also been described[115] which is designed to measure changes of the emission spectra with temperature. The photographic recording method, however, makes this spectrometer insensitive compared to other spectrometers.

VI. ASPECTS OF INSTRUMENTS FOR TL DATING

In archaeological dating by TL techniques (cf. Volume I, Chapter 3) the dominant instrumental problem is to achieve a sensitivity high enough for detection of the weak TL from archaeological samples. All the procedures previously described in this chapter to increase the signal-to-noise ratio must therefore be considered when designing an instrument for TL dating. Special attention must be paid to the problems imposed by the relatively high readout temperatures (approximately 500°C) that are often used in TL dating. Reduction of incandescent light emission and PM tube stabilization are, therefore, especially important. The predose TL dating technique, e.g., uses the 110°C glow peak of quartz and includes heating the quartz samples to 500°C. This annealing can be carried out either in the TL reader or in an external oven. If performed in the TL reader, it may be necessary to use a cooling system for the readout chamber to avoid excessive heating of the sample changer and the readout chamber as a result of successive heatings. A systematic error may otherwise be introduced by the increasing temperature since the 110°C glow peak in quartz has a relatively short half-life even at room temperature, about 3.5 hr at 110°C, and the fading rate increases with increasing temperature.

Samples used in TL dating often consist of small grains of quartz or feldspar. Since the grains have a relatively large surface-to-volume ratio, nonradiation-induced signals may also constitute a significant problem. As previously discussed, the main methods to reduce these signals are read out in an oxygen-free atmosphere (cf. Section IV.B.2), e.g., nitrogen, and the use of optical filters to selectively suppress the nonradiation-induced TL.[60] The purity of the inert gas is critical; the oxygen content must not be greater than a few parts per million. A nitrogen flow of several liters per minute is more efficient in reducing the nonradiation-induced signals than a static nitrogen atmosphere, probably because of desorption of oxygen from the sample during heating.[116]

The TL dating of an archaeological object requires many TL measurements on different samples so that manual sample changing is overly time consuming. Bøtter-Jensen and Mejdahl[117] developed an automated TL readout instrument especially designed to circumvent this problem. The samples are contained in 24 small platinum cups placed in cutouts on the periphery of a sample changer in the form of a turntable. When a sample is to be heated, a heating element is raised through the cutout lifting the platinum cup with the sample. The instrument contains a ^{90}Sr-^{90}Y beta source for irradiation of the samples. Two irradiation series, each followed by readout of the samples, are performed for the determination of the archaeological dose. The first irradiation series yields the so-called TL build-up curve, i.e., the TL as a function of added dose (cf. Volume III, Chapter 3), while the other is performed individually to calibrate the different samples in the 24 cups. The individual calibration allows sample mass corrections in a relatively easy way. Weighing errors during the pouring of the sample into the cups are therefore not critical. The whole irradiation and readout procedure is controlled by a microprocessor and takes about 2 hr.

The incorporation of the ^{90}Y-^{90}Sr radiation source in the readout unit increased the PM tube dark current.[118] The increase was caused by insufficient shielding of the radiation source so that radiation, probably X-rays generated during the slowing down of β particles, caused

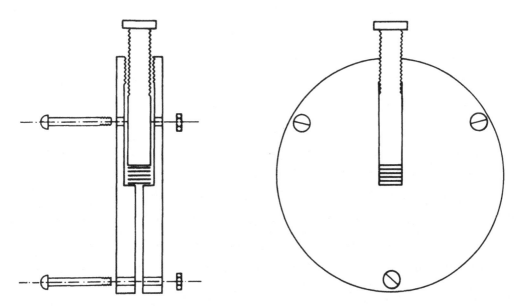

FIGURE 20. A teflon® phantom for homogeneous irradiation of LiF-teflon® disometers. (After Carlsson, C. A., Mårtensson, B. K. A., and Alm Carlsson, G., Proc. 2nd Int. Conf. Luminescence Dosimetry, U.S. A.E.C. CONF-680920, NTIS, Springfield, Va., 1968.)

light emission in the PM tube and the optical parts in the reader. Placing lead bricks between the radiation source and the PM tube decreased the dark current to its normal value.

VII. CALIBRATION PHANTOM

It is difficult to manufacture TL dosimeters with equal efficiencies, i.e., dosimeters emitting equal TL per unit absorbed dose in the dosimeter. Standard deviations around 5% for the individual sensitivities of the dosimeters in a group are not unusual. This is a larger spread than can be tolerated in many dosimetry measurements. One can then either select dosimeters with similar sensitivities from a large group of dosimeters in order to decrease the spread in the sensitivities or determine individual calibration constants for all dosimeters. Both methods require that the individual sensitivities of the dosimeters are determined by irradiating all dosimeters to the same absorbed dose. The precision that is needed when the individual sensitivities are assessed depends on the precision that is required in the dosimetric measurements. In high precision measurements it is generally easier to apply individual calibration constants than to select dosimeters with sufficiently narrow sensitivity distribution.

The individual calibration constants, c_i, are calculated[10] after read out of dosimeters irradiated to the same absorbed dose.

$$c_i = \frac{\frac{1}{n} \sum_{i=1}^{n} x_i}{x_i}$$

where x_i is the TL signal from dosimeter i and n is the number of dosimeters.

The calibration constants therefore show the variation of the sensitivities of the dosimeters around a common mean value. In order to perform high precision measurements all the dosimeters must be irradiated to the same absorbed dose during the determination of the individual calibration constants. This is facilitated by irradiating all dosimeters at the same time in a special calibration phantom, provided the radiation field is uniform. Carlsson et al.[10] have described a calibration phantom for LiF-teflon® dosimeters which may serve as a design model (Figure 20).

During the calibration irradiation the dosimeters should be tightly packed in a stack in the phantom and the direction of the radiation should be perpendicular to the stack axis. The phantom material should be similar in atomic composition and density to the dosimeter material. For LiF dosimeters teflon® is a good choice[10] when ^{137}Cs or ^{60}Co gamma rays are used as primary radiation. The similarity in atomic composition and density ensures that the photon and electron spectra in the phantom are independent of the number of dosimeters used. This minimizes the dose gradient along the dosimeter stack because of differences between the dosimeter and the phantom material. The phantom should be thick enough so that secondary particle build-up at the position of the dosimeters is complete, again, in order to minimize dose gradients in the dosimeters. By irradiating the phantom in two opposing directions perpendicular to its flat surfaces as suggested by Carlsson et al.,[10] the depth dose over dosimeters of diameters as large as 13 mm becomes flat. The dosimeters used by Carlsson et al.[10] were thin (0.1 mm) and 25 dosimeters occupied only 3 × 13 mm. Inhomogeneities in the radiation field could therefore be neglected. For thicker dosimeters field inhomogeneities may become a problem.

The use of a calibration phantom based on the same design principles as those described above facilitates an accurate determination of the individual calibration constants for the dosimeters and is almost a prerequisite in achieving optimal precision (σ = 0.2 to 0.3%) in TL measurements.[10,54]

VIII. ASPECTS OF ANNEALING OVENS

The TL efficiency is known to depend on thermal history in many materials. Thermal treatments before the irradiation are carried out to optimize the trap distribution in the material and to empty previously filled high temperature traps, while thermal treatments after the irradiation but before readout generally serve to eliminate TL peaks which fade during the measuring interval. It is often possible to carry out the necessary thermal treatments in the readout equipment prior to readout, however, in many cases it may be more efficient and rapid to use external ovens for the thermal treatments.

The first requirement of a TL annealing oven is a temperature control system sufficiently accurate to keep the temperature constant within narrow limits. Muffle ovens are known to possess this property but they often have another characteristic that makes them less favorable than ovens in which hot air circulated by a fan is utilized for heating. The temperature regulation in muffle ovens uses a thermocouple that protrudes from the oven walls. When a muffle oven is opened, even for a short time, the air in the oven is cooled since it is mixed with air at room temperature. The decrease in the air temperature is sensed by the thermocouple and in consequence the power to the heating elements in the stone walls of the oven is increased. However, the temperature of the oven walls does not significantly decrease during the opening because of their high thermal capacity. The increased power to the heating elements therefore causes a temperature overshoot when the oven is closed after insertion of the dosimeters.

The overshoot appears during the first 10 to 15 min after the closing and it may be as large as 20°C at 400°C. In ovens utilizing circulating hot air, overshoot problems are minimal since the temperature monitoring is made in the air jet and it is easy to increase the air temperature rapidly because of its low thermal capacity. Such ovens are therefore preferred over muffle ovens.

Dosimeters must often be stored in special containers during thermal treatments. Such a container may consist of a metal plate with depressions covered with another metal plate. It is important that the container has a high thermal conductivity to enable rapid heating and to minimize the temperature gradient. When reflective metal is used as a container, it is important to realize that the temperature in the oven may change from the nominal temperature because of the reflectivity. This is so because the reflective plate may reflect more thermal radiation towards the thermocouple than the oven walls and this increases the temperature of the thermocouple. As a result, the power to the heating elements decreases and the temperature at the dosimeter container also decreases. This effect is more pronounced for smaller ovens. A difference of 90°C between the nominal temperature of 400°C and the actual temperature has been observed in our laboratory[119] when an Al plate covered the bottom of a muffle oven.

Another very important factor that must be considered is the time required for the dosimeters to reach the temperature of the oven. How long this takes depends on the thermal conductivities of the dosimeters and their container, and on the temperature and the type of the oven. The time needed is in most cases several minutes, but may be as large as half an hour or more.[1] Many apparently contradictory results of thermal treatments performed by various authors can probably be explained by short thermal treatment times that did not allow the TL material to reach the oven temperature.

Not only the heating but also the cooling of dosimeters should be uniform and reproducible since the TL efficiency in many cases depends on the cooling rate. For example, reproducible cooling is a prerequisite for successful dosimetry with LiF:Mg,Ti dosimeters.

Ovens that are used for thermal treatments of TL dosimeters should be used solely for this purpose. Any other usage may contaminate the oven with material that can be transferred to the dosimeters during heating. The material may then emit nonradiation-induced TL during the readout of the contaminated dosimeters. Robertson[11] has reported that annealing of LiF-teflon® dosimeters in an oven which had been used to speed up the setting of Araldite® induced large nonradiation-induced TL signals.

REFERENCES

1. **McKinlay, A. F.,** *Thermoluminescence Dosimetry,* Adam Hilger Ltd., Bristol, 1981.
2. **Campion, P. J., Burns, J. E., and Williams, A.,** *A Code of Practice for the Detailed Statement of Accuracy,* National Physical Laboratory, London, 1973.
3. American National Standards Institute, Performance, Testing, and Procedural Specifications for Thermoluminescence Dosimetry (Environmental Applications), ANSI N454-1975, New York, 1975.
4. **Young, A. T.,** Photometric error analysis. IX. Optimum use of photomultipliers, *Appl. Opt.,* 8, 2431, 1969.
5. **Spanne, P.,** Thermoluminescence dosimetry in the μGy range, *Acta Radiol.,* Suppl. 360, 1979.
6. **Bjärngard, B. E. and Jones, D.,** Lithium fluoride-teflon thermoluminescence dosimeters, Proc. Int. Conf. Luminescence Dosimetry. U.S. A.E.C. CONF-650637, NTIS, Springfield, Va., 1967, 308.
7. **Carter, J. G., Birkhoff, R. D., and Nelson, D. R.,** Importance of thermal equilibrium in thermoluminescence measurements, *Health Phys.,* 10, 539, 1964.
8. **Burgkhardt, B. and Piesch, E.,** The effect of post irradiation annealing on the fading characteristics of different thermoluminescent materials. II. Optimal treatment and recommendations, *Nucl. Instrum. Methods,* 155, 299, 1978.
9. **Carlsson, C. A.,** Thermoluminescence of LiF: dependence of thermal history, *Phys. Med. Biol.,* 14, 107, 1969.
10. **Carlsson, C. A., Mårtensson, B. K. A., and Alm Carlsson, G.,** High precision dosimetry using thermoluminescent LiF, Proc. 2nd Int. Conf. Luminescence Dosimetry, U.S. A.E.C. CONF-680920, NTIS, Springfield, Va., 1968.
11. **Robertson, M. E. A.,** *Identification and Reduction of Errors in Thermoluminescence Dosimetry Systems,* D. A. Pitman Ltd., Weybridge, Surrey, 1975.
12. **Burch, W. H.,** Thermoluminescence, low radiation dosage and black-body radiation, *Phys. Med. Biol.,* 12, 523, 1967.

13. **Webb, G. A. M. and Phykitt, H. P.,** Possible elimination of the annealing cycle for thermoluminescent LiF, in Proc. 3rd Int. Conf. Luminescence Dosimetry, Rep. No. 249, IAEA/AEC, Risø, Denmark, 1971.
14. **Gorbics, S. G., Nash, A. E., and Attix, F. H.,** Thermal quenching of luminescence in six dosimetry phosphors. II. Quenching of thermoluminescence, Proc. 2nd Int. Conf. Luminescence Dosimetry, U.S. A.E.C. CONF-680920, NTIS, Springfield, Va., 1968.
15. **Perry, K. F. G. and Preston, H. E.,** Progress towards a thermoluminescent dosimetry system for large-scale personnel monitoring, presented at the 2nd Int. Congr. Int. Radiat. Prot. Soc., Brighton, England, May 3 to 8, 1970.
16. **Spanne, P.,** Warning against high energy β-emitters in light sources for TL-readers, *Health Phys.,* 24, 568, 1973.
17. **Hartin, W. J.,** An improved thermoluminescence dosimetry system, *Health Phys.,* 13, 567, 1967.
18. **Jones, A. R.,** A personal dosimeter system based on lithium fluoride thermoluminescent dosimeters (TLD), in Proc. 3rd Int. Conf. Luminescence Dosimetry, Rep. No. 249, IAEA/AEC, Risø, Denmark, 1971, 831.
19. **Cox, F. M. and Lucas, A. C.,** An automated thermoluminescent dosimetry (TLD) system for personnel monitoring. I. System description and preliminary results, *Health Phys.,* 27, 339, 1974.
20. **Portal, G., Blanchard, P., Prigent, R., and Chenault, R.,** Automatic rapid reading system for thermoluminescent personnel dosimeters, in Proc. 4th Int. Conf. Luminescence Dosimetry, Niewiadomski, T., Ed., Institute of Nuclear Physics, Krakow, 1974, 617.
21. **Shipler, D. B., Nichols, L. L., and Kocher, L. F.,** Thermoluminescence from adhesive coated teflon tape, *Health Phys.,* 22, 195, 1972.
22. **Grogan, D., Ashmore, J. P., Bradeley, R., and Scott, I. G.,** A national TLD service-operation and assessment, in Proc. 5th Int. Conf. Luminescence Dosimetry, Scharmann, E., Ed., Physikalisches Institut, Giessen, 1977, 207.
23. **Portal, G., Prigent, R., Blanchard, Ph., and Chenault, R.,** A new automatic reader for big scale routine TL personal dosimeters, in Proc. 5th Int. Conf. Luminescence Dosimetry, Scharmann, A., Ed., Physikalisches Institute, Giessen, 1977, 257.
24. **Petrock, K. F. and Jones, D. E.,** Hot nitrogen gas for heating thermoluminescent dosimeters, in Proc. 2nd Int. Conf. Luminescence Dosimetry, U.S. A.E.C. CONF-680920, NTIS, Springfield, Va., 1968.
25. **Bøtter-Jensen, L.,** A Reader for the Measurement of Solid Thermoluminescence Dosimeters by Means of Hot Nitrogen Gas, Rep. No. M-1238, Risø, Denmark, 1970.
26. **Bøtter-Jensen, L.,** Read out instrument for solid thermoluminescence dosimeters using hot nitrogen gas as heating medium, in Proc. Symp. Adv. Phys. Biol. Radiat. Detectors, SM-143/20, IAEA, Vienna, 1971.
27. **Bøtter-Jensen, L.,** A simple hot N_2-gas TL reader incorporating a post-irradiation annealing facility, *Nucl. Instrum. Methods,* 153, 413, 1978.
28. **Bøtter-Jensen, L. and Christensen, P.,** Progress towards automatic TLD processing for large scale routine monitoring at Risø, in Proc. 3rd Int. Conf. Luminescence Dosimetry, Rep. No. 249, IAEA/AEC, Risø, Denmark, 1971, 851.
29. **Julius, H. W., Verkoef, C. W., Busscher, F., and Oterman, F.,** A versatile automatic TLD system under development, in Proc. 4th Int. Conf Luminescence Dosimetry, Niewiadomski, T., Ed., Institute of Nuclear Physics, Krakow, 1974, 675.
30. **Julius, H. W.,** Instrumentation, in *Applied Thermoluminescence Dosimetry,* Oberhofer, M. and Scharmann, A., Eds., Adam Hilger Ltd., Bristol, 1981, chap. 3.
31. **Toivonen, M.,** Automated read-out of thermoluminescence dosimeters in a centralized individual monitoring service, in Proc. 5th Int. Conf. Luminescence Dosimetry, Scharmann, A., Ed., Physikalisches Institut, Giessen, 1977, 274.
32. **Toivonen, M.,** Individual TL Detector Characteristics in Automated Processing of Personnel Dosimeters: Correction Factors as Extension of Identification Codes of Dosimeter Cards, Rep. No. STL-A27, Institute of Radiation Protection, Helsinki, 1979.
33. **Oonishi, H., Yamamoto, O., Yamashita, T., and Hasegawa, S.,** Dosimeter and reader by hot air jet, in Proc. 3rd Int. Conf. Luminescence Dosimetry, Rep. No. 249, Mejdahl, V., Ed., IAEA/AEC, Risø, Denmark, 1971, 237.
34. **Hasegawa, S., Iga, K., Miyagawa, K., Oktsuka, H., Kunishige, H., and Nakao, F.,** A large scale TLD system using data processing for radiation monitoring, in Proc. 4th Int. Conf. Luminescence Dosimetry, Niewiadomski, T., Ed., Institute of Nuclear Physics, Krakow, 1974, 627.
35. **Widell, C.-O.,** personal communication, 1979.
36. **Spanne, P.,** Spurious signals from TL dosimeters: comparison between hot gas and linear heating systems, presented at 6th Meet. Nordic Soc. Radiat. Prot., Reykjavik Island, June 18 to 20, 1981.
37. **Cameron, J. E., Zimmerman, D. W., Bland, R. W., Suntharalingam, N., and Kenney, K. N.,** *Thermoluminescent Dosimetry,* The University of Wisconsin Press, Madison, 1968, 92.
38. **Bailiff, I. K. and Aitken, M. J.,** Use of thermoluminescence dosimetry for evaluation of internal beta dose-rate in archeological dating, *Nucl. Instrum. Methods,* 173, 1980, 423.

39. **Yamamoto, O., Yasuno, Y., Minamide, S., Hasegawa, S., Tsutsui, H., Takenaga, M., and Yamashita, T.,** Construction of a composite thin-element TLD using an optical heating method, *Health Phys.,* 43, 383, 1982.

40. **Brunskill, R. T. and Langmead, W. A.,** A sealed thermoluminescent dosimeter employing R.F. heating for routine individual monitoring, in Proc. Adv. Phys. Biol. Radiat. Detectors, SM-143, IAEA, Vienna, 1971, 67.

41. **Schulman, J. H.,** Survey of luminescence dosimetry, in Proc. Int. Conf. Luminescence Dosimetry, U.S. A.E.C. CONF-650637, NTIS, Springfield, Va., 1967, 3.

42. **Benner, S., Johansson, J. M., Lindskoug, B., and Nyman, P. T.,** A miniature LiF dosimeter for in vivo measurements, in Proc. Solid State and chemical radiation dosimetry in medicine and biology, SM78/25, IAEA, Vienna, 1967, 65.

43. **Lindskoug, B., Johansson, M., Karlsson, R., and Kjellgren, R.,** Measuring device for thermoluminescence dosimetry, *J. Sci. Instrum.,* 44, 939, 1967.

44. **Lindskoug, B. and Bengtsson, B.-E.,** Automated thermoluminescence reader. I. Technical construction and function, *Acta Radiol. Ther. Phys. Biol.,* 14, 195, 1975.

45. **Lindskoug, B.,** Automated thermoluminescence reader. II. Experiments and theory, *Acta Radiol. Ther. Phys. Biol.,* 14, 347, 1975.

46. **Lindskoug, B. A.,** Further consideration on the use of continuous cylindrical dosimeters in TSL dosimetry, *Nucl. Instrum. Methods,* 175, 89, 1980.

47. **Carlsson, C. A. and Alm Carlsson, G.,** Induced LET-dependence in thermoluminescent LiF and its application as LET-meter, Proc. 2nd Int. Conf. Luminescence Dosimetry, U.S. A.E.C. CONF-680920, NTIS, Springfield, Va., 1968.

48. **Suntharalingam, N. and Cameron, J. R.,** Thermoluminescent response of lithium fluoride to radiations with different LET, *Phys. Med. Biol.,* 14, 397, 1969.

49. **Bron, R. and Valladas, G.,** Appareil pour la mesure de la thermoluminescence de petits énchantillons, *Nucl. Instrum. Methods,* 127, 109, 1975.

50. **Grogan, D., Ashmore, J. P., and Bradeley, R. P.,** Centralized TLD service and record keeping in Canada, IAEA Int. Symp. Adv. Radiat. Prot. Monitoring, SM-229/138, IAEA, Vienna, 1978.

51. **Karzmark, C. J., Fowler, J. F., and White, J. T.,** Problems of reader design and measurement Int. Conf. Luminescence Dosimetry, U.S. A.E.C. CONF-650637, 1967, 265.

52. International Atomic Energy Agency, IAEA laboratory activities, STI/DOC/10/55, Vienna, 1965.

53. **Regulla, D. F.,** Material handling and measuring techniques in thermoluminescence dosimetry, presented at CEC Course: Thermoluminescence Dosimetry, TD/77/8, CEC, Ispra, 1977.

54. **Alm Carlsson, G.,** Dosimetry at interfaces. Theoretical analysis and measurements by means of thermoluminescent LiF, *Acta Radiol.,* Suppl. 332, 1973.

55. **Lakshmanan, A. R. and Bhatt, R. C.,** Low dose measurements with $CaSO_4$:Dy teflon dosimeters, *Int. J. Appl. Radiat. Isot.,* 33, 707, 1982.

56. **Tutiya, M.,** White residue released from irradiated polytetrafluorethylene by heat treatment, *Jpn. J. Appl. Phys.,* 8, 1356, 1969.

57. **Nash, A. E. and Johnson, T. L.,** LiF (TLD-600) thermoluminescence detectors for mixed thermal neutron and gamma dosimetry, in Proc. 5th Int. Conf. Luminescence Dosimetry, Scharmann, A., Ed., Physikalisches Institut, Giessen, 1977, 393.

58. **Webb, G. A. M.,** The measurement of integrated gamma ray doses in the range 50 mrad to 5 mrad using phosphate glass and lithium fluoride, in Proc. Symp. Personnel Dosimetry for Radiation Accidents, IAEA, Vienna, 1965, 149.

59. **Nash, A. E., Johnson, T. L., Attix, F. H., and Schulman, J. H.,** Spurious thermoluminescence of CaF_2:Mn and LiF (TLD-100), in Proc. Int. Conf. Luminescence Dosimetry, U.S. A.E.C. CONF-650637, NTIS, Springfield, Va., 1967, 244

60. **Fleming, S. J.,** The colour of spurious thermoluminescence in dosimetry phosphors, in Proc. 2nd Int. Conf. Lumincescence Dosimetry, U.S. A.E.C. CONF-680920, NTIS, Springfield, Va., 1968, 266.

61. **Lingertat, J.,** Untersuchungen zur Personendosimetrie mit Lithiumfluorid, Ph.D. thesis, Technische Universität, Dresden, 1967.

62. **McCall, R. C. and Fix, R. C.,** A sensitive LiF dosimeter for routine beta and gamma personnel monitoring, *Health Phys.,* 10, 602, 1964.

63. **Kastner, J., Hukko, R., and Oltman, B. G.,** Thermoluminescent dosimetry for beta rays, in Proc. Int. Conf. Luminescence Dosimetry, U.S. A.E.C. CONF-650637, NTIS, Springfield, Va., 1967, 482.

64. **Aitken, M. J., Alldred, J. C., Thompson, J., Reid, J., Tite, M. S., and Fleming, S. J.,** Quenching of spurious thermoluminescence by nitrogen, in Proc. Int. Conf. Luminescence Dosimetry, U.S. A.E.C. CONF-650637, NTIS, Springfield, Va., 1967, 236.

65. **Svarcer, V. and Fowler, J. F.,** Spurious thermoluminescence and tribothermoluminescence in lithium fluoride dosimetry powder, in Proc. Int. Conf. Luminescence Dosimetry, U.S. A.E.C. CONF-650637, NTIS, Springfield, Va., 1967, 227.

66. **Kathuria, S. P., Sunta, C. M., Sasidharan, R., and Jain, V. K.,** Spurious thermoluminescence in TL phosphors, in Proc. Natl. Symp. Thermoluminescence and Its Applications, Madras, 1976, 680.

67. **Carlsson, R. W. and Rickey, J. B.,** The Harshaw model 2000 thermoluminescence analyzer, in Proc. 2nd Int. Conf. Luminescence Dosimetry, U.S. A.E.C. CONF-680920, NTIS, Springfield, Va., 1968, 706.

68. **Prescott, J. R.,** A statistical model for photomultiplier single-electron statistics, *Nucl. Instrum. Methods,* 39, 173, 1966.

69. **Murray, R. B. and Manning, J. J.,** Response of end window photomultiplier tubes as a function of temperature, *IRE Trans. Nucl. Sci.,* NS-7, 80, 1960.

70. **Boileau, A. R. and Miller, F. D.,** Changes in spectral sensitivity of multiplier phototubes resulting from changes in temperature, *Appl. Opt.,* 6, 1179, 1967.

71. **Budde, W. and Kelly, P.,** Variations of the spectral sensitivity of RCA 6217 and 5819 photomultipliers at low temperature, *Appl. Opt.,* 10, 2612, 1971.

72. **Saunders, J. E.,** Significant changes in TLD readings produced by AC heater currents, in Proc. 3rd Int. Conf. Luminescence Dosimetry, Rep. No. 249, Mejdahl, V., Ed., IAEA/AEC, Risφ, Denmark, 1971, 209.

73. **Forslo, H., Carlsson, C. A., Spanne, P., and Wik, S.,** PM-tube sensitivity as a function of heating current in thermoluminescence dosimetry, in Proc. 4th Int. Conf. Luminescence Dosimetry, Niewiadomski, T., Ed., Institute of Nuclear Physics, Krakow, 1974, 724.

74. **Coates, P. B.,** The edge effect in electron multiplier statistics, *J. Phys. D,* 5, 1972, 915.

75. **Eberhardt, E. H.,** Threshold sensitivity and noise ratings of multiplier phototubes, *Appl. Opt.,* 6, 251, 1967.

76. **Rodman, J. P. and Smith, H. J.,** Tests of photomultipliers for astronomical pulse counting applications, *Appl. Opt.,* 2, 181, 1963.

77. **Barton, J. C., Barnaby, C. F., and Jasani, B. M.,** An investigation of noise in venetian blind photomultipliers, *J. Sci. Instrum.,* 41, 599, 1964.

78. **Gadsden, M.,** Some statistical properties of pulses from photomultipliers, *Appl. Opt.,* 4, 1446, 1965.

79. **Oliver, C. J. and Pike, E. R.,** Measurement of low light flux by photon counting, *Br. J. Appl. Phys.,* 1, 1459, 1968.

80. **Harker, Y. D., Masso, J. D., and Edwards, D. F.,** Merits of photomultiplier cooling for photon experiments, *Appl. Opt.,* 8, 2563, 1969.

81. **Zagorites, H. A. and Lee, Y. D.,** Gamma and X-ray effects in multiplier phototubes, *IEEE Trans. Nucl. Sci.,* NS-12, 343, 1964.

82. **Young, A. T.,** Cosmic ray induced dark current in photomultipliers, *Rev. Sci. Instrum.,* 37, 1472, 1966.

83. **Dressler, K. and Spitzer, L., Jr.,** Photomultiplier tube pulses induced by γ-rays, *Rev. Sci. Instrum.,* 38, 436, 1967.

84. **Jerde, R. L., Petersen, L. E., and Stein, W.,** Effects of high energy radiations on noise pulses from photomultiplier tubes, *Rev. Sci. Instrum.,* 38, 1387, 1967.

85. **Johnson, S. M., Jr.,** Radiation effects on multiplier phototubes, in Proc. Nucl. Sci. Symp., Miami, 1972, 1.

86. **Young, A. T.,** Photometric error analysis IX. Optimum use of photomultipliers, *Appl. Opt.,* 8, 2431, 1969.

87. **Clay, R. W. and Gregory, A. G.,** Photomultiplier noise associated with cosmic rays, *J. Phys. A,* 10, 135, 1977.

88. **Coates, P. B.,** The origins of afterpulses in photomultipliers and prepulse height distribution, *J. Phys. D,* 6, 1862, 1973.

89. **Schneider, R. M., Kirk, W. P., Steiner, J. F., Jr., and Rechen, H. J. L.,** Photon counter for the measurement of thermoluminescence, *Rev. Sci. Instrum.,* 39, 1369, 1968.

90. **Aitken, M. J., Alldred, J. C., and Thompson, J.,** A photon ratemeter system for low level thermoluminescence measurements, in Proc. 2nd Int. Conf. Luminescence Dosimetry, U.S. A.E.C. CONF-680920, NTIS, Springfield, Va., 1968, 248.

91. **Schlesinger, T., Avni, A., and Feige, Y.,** Photon counting as applied to thermoluminescent dosimetry, in Proc. 3rd Int. Conf. Luminescence Dosimetry, Rep. No. 249, Mejdahl, V., Ed., IAEA, AEC, Risφ, Denmark, 1971, 226.

92. **Niewiadomski, T.,** Determination of optimum conditions for photon counting in thermoluminescence measurements, SM-160, IAEA, Vienna, 1972, 199.

93. **Lasky, J. B., Pearson, D. W., and Moran, P. R.,** A photon counting system for thermoluminescence dosimetry, U.S. A.E.C.-C00-1105-188, NTIS, Springfield, Va., 1973.

94. **Owaki, S., Yamauchi, M., and Kawanishi, M.,** Low dose measurements with TLD by single photon counting, in Proc. 5th Int. Conf. Luminescence Dosimetry, Scharmann, A., Ed., Physikalisches Institut, Giessen, 1977, 183.

95. **Becker, K.,** *Solid State Dosimetry,* CRC Press, Boca Raton, Fla., 1973, 85.

96. **Lippert, J. and Mejdahl, V.**, Thermoluminescent readout instrument for measurement of small doses, in Proc. Int. Conf. Luminescence Dosimetry, U.S. A.E.C. CONF-650637, NTIS, Springfield, Va., 1967, 204.

97. **Horowitz, Y. S.**, The theoretical and microdosimetric basis of thermoluminescence and applications to dosimetry, *Phys. Med. Biol.*, 26, 765, 1981.

98. **Spanne, P.**, Selecting PM tube power supply voltages for TLD readers, *Nucl. Instrum. Methods*, 175, 1980, 92.

99. **Kellerer, A. M.**, Mikrodosimetrie. Grundlagen einer Theorie der Strahlenqualität, GSF-Bericht B-1, Gesellschaft für Strahlenforschung M.B.H., München, 1968.

100. **Au, Y.-E. and Moran, P. R.**, Evaluation of the UDT-500 photodiode-operational amplifier pair as detector for thermoluminescence, U.S. A.E.C.-C00-1105-177, University of Wisconsin, Madison, 1972.

101. **Cova, S., Longoni, A., and Andreoni, A.**, Towards picosecond resolution with single-photon avalanche diodes, *Rev. Sci. Instrum.*, 52, 1981, 408.

102. **Burghkhardt, B. and Piesch, E.**, Systematical and statistical errors in using reference light sources to calibrate TLD readers, *Health Phys.*, 40, 549, 1981.

103. **Reiter, W. L. and Stengl, G.**, A long term stable reference light source using LEDs for stabilization of scintillation spectrometers, *Nucl. Instrum. Methods*, 173, 1980, 275.

104. **Reiter, W. L. and Stengl, G.**, A stabilizing system for scintillation spectromers with a light emitting diode and a PIN photodiode, *Nucl. Instrum. Methods*, 169, 1980, 469.

105. **Sanharan, A. and Kannan, S.**, A stable and low cost light emitting diode light source for TLD reader calibration, in Proc. Natl. Symp. Thermoluminescence Appl., Bhabha Atomic Research Centre, Bombay, 1975, 553.

106. **Gangadharan, P., Sanharan, A., and Kannan, S.**, A semi-automatic TLD personnel monitoring badge processor for routine use, in Proc. Natl. Symp. Thermoluminescence Appl., Bhabha Atomic Research Centre, Bombay, 1975, 563.

107. **Harris, A. M. and Jackson, J. H.**, A rapid scanning spectrometer for the region 200-850 nm: application to thermoluminescent emission spectra, *J. Phys. E*, 3, 1970, 374.

108. **Harris, A. M. and Jackson, J. H.**, The emission of thermoluminescent dosimetry grade lithium fluoride, *J. Phys. D*, 3, 1970, 624.

109. **Strash, A. M. and Madey, R.**, The thermoluminescence spectrum and the energy conversion efficiency of lithium fluoride, in Proc. 2nd Int. Conf. Luminescence Dosimetry, U.S. A.E.C. CONF-680920, NTIS, Springfield, Va., 1968, 607.

110. **Oltman, B. G., Kastner, J., and Paden, C.**, Spectral analysis of thermoluminescence curves, in Proc. 2nd Int. Symp. Luminescence Dosimetry, U.S. A.E.C. CONF-680920, NTIS, Springfield, Va., 1968, 623.

111. **Fairchild, R. G., Mattern, P. L., Lengweiler, K., and Levy, P. W.**, Thermoluminescence of LiF TLD 100 dosimeter crystals, *IEEE Trans. Nucl. Sci.*, NS-21, 1974, 366.

112. **Fairchild, R. G., Mattern, P. L., Lengweiler, K., and Levy, P. W.**, Thermoluminescence of LiF TLD 100: emission-spectra measurements, *J. Appl. Phys.*, 49, 4512, 1978.

113. **Bailiff, I. K., Morris, D. A., and Aitken, M. J.**, A rapid-scanning interference spectrometer: application to low-level thermoluminescence emission, *J. Phys. E*, 10, 1977, 1156.

114. **McCall, R. C. and Ross, D. K.**, A method of measuring emission spectra from thermoluminescent materials, *Rev. Sci. Instrum.*, 40, 1969, 363.

115. **Beaven, H. J.**, An apparatus for recording changes with temperature of thermoluminescent glow spectra, *J. Phys. E*, 7, 1974, 708.

116. **Aitken, M. J. and Fleming, S. J.**, Thermoluminescence dosimetry in archaeological dating, in *Topics in Radiation Dosimetry, Radiation Dosimetry Suppl. 1*, Attix, F. H., Ed., Academic Press, London, 1972, 1.

117. **Bøtter-Jensen, L. B. and Mejdahl, V.**, Determination of archeological doses for TL dating using an automated apparatus, *Nucl. Instrum. Methods*, 175, 1980, 213.

118. **Bell, W. T., Mejdahl, V., and Winther-Nielsen, M.**, The refinement of the automated TL dating procedure and perspectives for the archeological application of the method as demonstrated by the results from sites of known age, in Proc. 20th Int. Symp. Archaeometry and Archaeological Prospection, Paris, March 14 to 17, 1980.

119. **Lindborg, L.**, Undersökning av några parametrar, som påverkar precision i dosmätning med termoluminescent LiF, Report Radiation Physics Department, University of Lund, 1970.

Chapter 2

CLINICAL APPLICATIONS OF THERMOLUMINESCENT DOSIMETRY

B. A. Lindskoug and L-M. Lundberg

TABLE OF CONTENTS

I. INTRODUCTION

The use of ionizing radiation in clinical practise must always be accompanied by the determination of the absorbed dose in tissue and there is also a growing realization of the importance of the development of appropriate methods for the checking of dose distributions. The use of radiation sources in clinical practice is constantly growing so that there is an ever-increasing need for dosimetry, particularly in diagnostic radiology where the greatest number of people are exposed to ionizing radiation.

The main need of dosimetry in medical radiology, which includes diagnostic radiology and radiation therapy, is in the mass-screening of the absorbed dose delivered in everyday routine work rather than in sophisticated scientific dosimetric investigations. The primary goal must be to ensure that the dose limits recommended by radiation protection are not exceeded and, in therapy, that the expected dose is reached within acceptable limits. Overdoses may be harmful to the patient and underdoses do not achieve optimal cure. Besides monitoring the radiation load to patients and personnel, dose measurements can also test equipment quality and irradiation procedures, e.g., the beam parameters of a computerized tomography scanner or a therapy unit.

Thermoluminescent dosimetry (TLD) is used both in diagnostic radiology and in radiation therapy. A number of publications can be cited which describe successful in vivo and in vitro clinical dose measurements.[1-10] This chapter deals with absorbed dose determination by means of TLD in medical radiology. Note that the word "dose" shall always be interpreted as the "absorbed dose", and if no specific material is mentioned it is tacitly understood that reference is made to the absorbed dose in tissue, or in some tissue-equivalent material, like water or polystyrene.

II. HISTORY

The history of clinical dosimetry in the Nordic countries goes back to the pioneering work of Rolf Sieverts, who established and organized the control of medical radiology in the 1920s. He described, in 1932, a condenser chamber with dimensions small enough to be used for in vivo clinical dosimetry.[11] Other physicists continued this work by developing further the condenser chamber and its practical use within clinical dosimetry.[12-20] Ionization

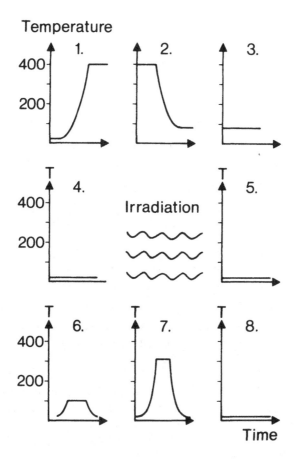

FIGURE 1.   The thermal cycle of a TL dosimeter going through the standard annealing procedure and readout.[35] (1) High temperature, preirradiation annealing (400°C, 1 hr). (2) Critical cooling. (3) Low temperature annealing (80°C, 24 hr). (4) Storage in room temperature. (5) Irradiation at room temperature. (6) Eventually low temperature postirradiation annealing (100°C, 10 min). (7) Readout heating, and critical cooling to room temperature. (8) Storage in room temperature.

chambers are, however, strongly dependent on environmental factors, such as air pressure, temperature, and humidity. They are also mechanically fragile. These properties are serious drawbacks in clinical work, which expedited the search for alternative methods. Interest in TL dosimetry started to grow in the Nordic countries when Daniels and colleagues introduced it in the late 1940s as a practical method for clinical dosimetry.[21] The condenser ionization chambers were gradually replaced in the 1960s by solid-state dosimeters of which the TL dosimeters held a dominant position due to their extremely small dimensions, independence of environmental factors, and mechanical stability. The TL dosimeter is everything that the ionization chamber is not. A few drawbacks do exist, however. TL materials are usually sensitive to ultraviolet light which may introduce disturbances at low dose (increased background) and some materials, like $Li_2B_4O_7$, can be influenced by humidity due to their hygroscopicity.

Dosimetric TL research continued to enhance the advantages of the new TL techniques in clinical dosimetry which gradually developed into a contemporary method of dosimetric TL measurements.[22-28] Lindskoug and Bengtsson[29] developed a TL dosimetry system with a high degree of automation coupled with simple handling of the dosimeters which was

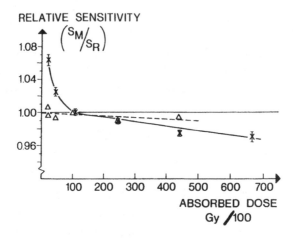

FIGURE 2.   The relative sensitivity, normalized to a reference readout with a constant dose (1 Gy), for dosimeters that are repeatedly used at various dose levels. The triangles ($\triangle$ — $\triangle$) correspond to two readouts with the residual signal subtraction.

particularly well suited for in vivo measurements. During this time of high activity in the field, Cameron et al[30] published the book that for many years was the reference work for thermoluminescent dosimetry. In 1973 Becker summarized much of the knowledge at that time concerning TL materials and methods[31] and a handbook, with a special emphasis on the applications of TLD, was published by McKinley.[32] Recently the ISPRA courses on TLD were published as a collection of lectures on applied thermoluminescence dosimetry[33] and finally, Horowitz recently summarized and discussed the literature and outstanding problems on the subject of the theoretical and microdosimetric basis of thermoluminescence and applications to dosimetry.[34]

III. HANDLING, ANNEALING, AND CALIBRATION IN A CLINICAL ENVIRONMENT

A. Handling

The handling of dosimeters in a clinical environment involves special considerations that are not normally encountered in a physics laboratory. The dosimeters are applied to the body of patients, either on the skin, or intracavitary, even interstitial and intravasal. This means that, for practical reasons, personnel other than physicists will handle the dosimeters and keep record of and care for the dosimeters. Consequently the staff must be fully informed about the applications and understand the meaning of the procedures involved and the results recorded. An average-size oncology center, treating about 1500 patients annually, will have to make approximately 150 routine readouts daily in order to keep a good checking record. Usually the same medical physics laboratory also gives TLD service to diagnostic radiology and nuclear medicine. Obviously, then, a firm routine must be established involving written instructions, simple identification of individual dosimeters, and facile collection and readout procedures.

Dirt and dust are enemies to all luminescence dosimetry. It is therefore necessary to work in a clean environment and to avoid directly marking the dosimeters. Vacuum forceps must always be used when handling the dosimeters and in the clinic they should be covered by some type of protective envelope.

B. Annealing

The most crucial parts of the procedures are the annealing and calibration of the dosimeters.

These must be the responsibilities of the physicists. The standard annealing procedure suggested by Zimmerman et al.[35] involves 400°C for 1 hr and 80°C for 24 hr. The thermal cycle of the dosimeters is schematically illustrated in Figure 1. Notice that the standard annealing procedure implies that the TL material, on two occasions, is cooled from a high temperature to room temperature, namely after readout and after the high temperature annealing. Precautions must be taken to ensure reproducible cooling because the cooling rate affects the TL efficiency of most dosimetric TL materials. The standard annealing method was initially developed for LiF crystals. Other TL materials are better treated with different annealing procedures. Teflon®-based dosimeters, e.g., cannot undergo the 400°C annealing since teflon® softens at about 350°C. Many attempts have been made to use TL materials without annealing.[27,36,37] Martensson[27] suggested that the reproducibility might be improved by excluding the high temperature annealing, but this is true only if the temperature gradient during cooling cannot be reproducibly controlled.

Using a nonanneal method, Lindskoug[37] found that the sensitization effect that occurs after fractionated irradiation and readout could be used to advantage. This method has been clinically tested for a decade with an ever increasing number of yearly measurements. The nonannealed TL materials, however, demonstrate increased sensitivity, particularly obvious in the dose range below 0.5 Gy, which depends among other factors on a residual signal which adds to the TL response. This residual effect can be eliminated by making two consecutive readouts and then subtracting the residual signal from the first measured response.[38] Figure 2 shows the sensitivity (S_M) for repeatedly used LiF dosimeters at various dose levels relative to the reference irradiation sensitivity (S_R), at a constant dose, with and without residual signal subtraction.

C. Calibration

Generally, the nonuniversality of the TL phenomenon makes it necessary to calibrate the dosimeters independent of the annealing method that is being used.[39] Furthermore, the LET dependence imposes the requirement of separate dosimeter calibration for each one of the radiation sources used in the laboratory. Finally, the fact that the LET dependence is also changing with accumulated dose makes repeated individual calibration necessary.[40] TL dosimeters are thus scondary dosimeters that cannot be used for the initial calibration of a radiation source, at least not at the present state of the art. Other methods, like using a primary standard ionization chamber, are better for the absolute calibration of the radiation sources. Even relative TL measurements may be subject to significant errors if the dosimeters are not properly standardized. In summary, each single TL dosimeter must be repeatedly calibrated for all the radiation sources intended for measurements and under all conditions, such as various depths, field sizes, filters, and blockings. At present this is the only way of using TL materials in clinical dosimetry if a high precision is required, i.e., a precision comparable to the substandard instruments. These extensive and repeated calibrations do, however, have an advantage in that they bypass many of the still unclarified problems of cavity theory in this field.[41-43] (See Volume II, Chapter 2, Section III.)

To be of use in clinical TL dosimetry all sources used in radiotherapy must be initially calibrated and the relative dose distribution in a water phantom must be determined under standard laboratory conditions. This means that the dose is known relative to an existing monitor system and a TL dosimeter can then be calibrated by placing it in a water-equivalent phantom exposed to a suitable calibration dose. Under nonstandard conditions, or in a diagnostic X-ray beam, the substandard dosimetry equipment is available for direct calibration in the beam.

The basic principle of TL dosimeter calibration is to determine a calibration factor that gives the absorbed dose to tissue, at the point of measurement, when the dosimeter is removed and replaced by tissue. If a dosimeter is to be calibrated for surface dose measurements, the calibration factor should give the dose to the surface of the tissue when the dosimeter is removed. It is necessary that the relative surface dose is known beforehand, determined by other methods such as a pancake, thin-walled ion chamber, or a small liquid ionization chamber.

If the dose at some depth is of interest, the dosimeter should preferably be calibrated at the relevant depth, or else on the surface with build-up material. It is not necessary that the build-up material have a thickness equal to d_{max} since this would be very impractical for high energy photons that may have a maximum depth of several centimeters. Usually, 5 mm of a tissue-equivalent material is sufficient, provided that the measurements and the calibration are carried out in the same geometry at all times.

The patient load on the treatment units often makes it troublesome to undertake repeated calibration of all the dosimeters for all sources. It may therefore be practical to dedicate one of the sources as a reference source. Reference irradiators containing ^{90}Sr-^{90}Y are commercially available. They can be used to advantage for high energy electrons and photons and also within some limited accumulated dose range for lower energies. This is because the accumulated dose-dependent sensitivity change will be different for sources with different average LET. Thus if there is a significant difference in LET between the reference source and some of the other sources, it is necessary to recalibrate after a certain limit of accumulated dose. Obviously, in a laboratory with a large variety of different sources, it is necessary to keep a computerized record of all the calibration factors for all the individual dosimeters. The detailed calculations involved in calibration and measurement are presented in Appendix 1.

A common method in TL dosimetry is to use some type of grouping of the dosimeters. Grouping means that the dosimeters are divided into matched groups with the same sensitivity, for instance with a maximum deviation of $\pm 2\%$. Some of the dosimeters are then calibrated and their calibration factor is used for all dosimeters within the group, i.e., no individual calibrations are necessary. In this way considerable calibration work is saved, however, it is necessary that all the dosimeters within the group be used in parallel at all times and be taken through the same thermal treatment and be given approximately the same accumulated dose.

It is observed that the random variation of the average response of a group of dosimeters is less than the variation of a single dosimeter response. Consequently, the random variation can be reduced by applying a single-detector correction factor that relates the sensitivity of a single dosimeter to the group average sensitivity.[44] The precision of measurement can be increased in this way. The use of relative single-detector correction factors is described in Appendix 2.

D. Calibration of Continuous Dosimeters

The continuous dosimeters, read out by linear motion through a heated oven, must be separately dealt with because they include a new parameter, the integration interval.[45] The integration interval can be set to any length between 1 and 100 mm, and there are only practical limits to the length of a dosimeter. The sensitivity may vary considerably along the length of a dosimeter so that each integration interval must be calibrated individually. The best method is to divide the response of an interval by its length which gives the calibration factors in units of Gy mm per TL response. This technique of measurement yields an enormous amount of data, particularly if the integration intervals are short. It is therefore necessary to link the readout instrument on-line to a computer used for data reduction and evaluation. A new calibration data file is generated on the mass-storage memory for each dosimeter and for each integration interval length used. This gives the freedom to choose any integration interval during measurement readout.

A phantom block, constructed of polystyrene slabs and suitable for the calibration of

FIGURE 3. Polystyrene phantom slabs for calibration of continuous dosimeters or measuring probes. A continuous dosimeter is inserted into the 2-mm spiral track.

continuous dosimeters or measuring probes, is shown in Figure 3. One of the slabs has a spiral track into which the dosimeters are inserted during exposure. The slab can be pressed between the other polystyrene slabs for calibration at any depth. The phantom is exposed to a large field that gives a sufficient flatness at the depth of reference. The calculations for calibration and measurements using continuous dosimeters are described in detail in Appendix 3.

IV. APPLICATIONS IN RADIOTHERAPY

A. Introduction

Although sophisticated computerized treatment planning systems are used in contemporary radiation therapy, the possibility of error cannot be excluded. No planning system is good enough to give a complete three-dimensional picture of the isodose distribution in the individual treatment situation. Besides, the routine use of computer planning can easily lead to unintended misuse if the computer data, or the algorithms, are used beyond the limits of their applications. The experimental primary data are never so complete that all individual patients can be satisfactorily planned. Thus cancer therapy by means of ionizing radiation introduces clinical situations where direct monitoring is necessary. The first parts of this section deal with electron and photon radiation only. A special section is devoted to neutrons.

The applications of TL dosimetry in radiation therapy may be divided into five main groups according to the specific application. These are

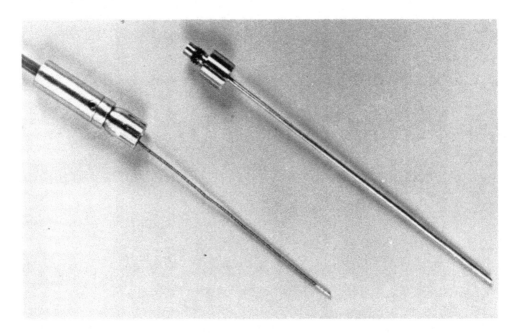

FIGURE 4. Right: a 10-cm implant needle for computerized afterloading. Left: a wire containing a sealed ^{192}Ir source (0.5 × 2 mm) at the tip. (Isotopentechnik, Dr. Sauerwein.)

1.   In vivo measurements in the region of interest
2.   In vivo measurements at some convenient region of the body, the data later corrected for appliction to the region of interest
3.   In vitro measurements such as in cadavers and autopsy specimens
4.   Measurements in anatomical phantoms
5.   Special phantom block studies, the result later checked under more realistic circumstances

Examples of applications are given in the order listed above.

B. General Verification of Computer Therapy Planning

The importance for in vivo measurements during radiotherapy arises from the need to check the absorbed dose distribution in complex treatment situations, in regions where anatomical structures are rapidly changing, and where irregular heterogeneities occur. These, in fact, include all parts of the body and most of the treatment situations.

Radiation therapy of a patient can go on for months. Not unusual are 20 and even 30 sessions. Blocks, wedges, filters, molds a.s.o. are used in order to improve the therapy. The dose distributions are generated by computer calculations, but very rarely more than one or two plans are produced for each patient in routine practice. The practical treatment, however, is always a volume concept. The paramount question, whether the dose distribution given to the patient really was the same as was planned, can be answered only by direct measurements on or inside the body of the patient during a therapy session. The absorbed dose must also be measured in the third dimension for which no planning was made. At least the first few sessions are monitored by direct measurements and all new treatment techniques are thoroughly investigated on site.

In vivo measurements are most frequently made on the skin of the patient's body. The purpose may be to check the skin dose, or to check the dose at some depth by applying known depth dose curves. For skin dose measurements, the dosimeters are thin and used without build-up material. If the absorbed dose at a certain depth is of concern, the dosimeters

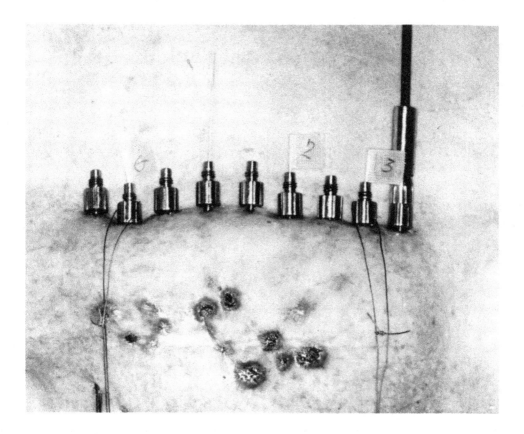

FIGURE 5.   Nine-needle surface implant for Ir-boost treatment of ca mammae. Four LiF-teflon® long-rod dosimeters are inserted into the needles, one of them pulled out halfway.

are covered by build-up material which usually does not have to be of full build-up thickness as long as the calibration is made under identical conditions (5-mm tissue-equivalent material is recommended).

Bagne defined an empirical function called the surface maximum ratio (SMR) in order to simplify the calculations in routine work.[46] SMR is the energy- and field size-dependent ratio of the TL response on the surface and at d_{max} in a tissue-equivalent phantom block. It is tabulated for all sources and field sizes. The dose at any depth can then be determined from a surface measurement at the entrance point of the central ray if the percentage depth dose is known. Corrections will be necessary at other positions in the portal. The empirical function SMR is described in detail in Appendix 4.

C. In Vivo Measurements in the Region of Interest

The region of interest for in vivo measurements is usually the target volume, but it may also be any adjacent organ.

The absorbed dose in superficial therapy can easily be checked by inserting dosimeters through needles implanted in the target volume. This technique of measurement is particularly well suited for use in interstitial therapy where needles or plastic tubes already perforate the target volume. If necessary, some extra needles may be implanted for the dosimetry. A most intriguing application of this method appears in the computerized interstitial afterloading technique.[47,48] In short, this technique involves optimization of the treatment by using a computer-guided source (^{192}Ir) that moves over predetermined intervals inside the implanted needles (Figure 4). The exposure times at the source stopping points are optimized for the

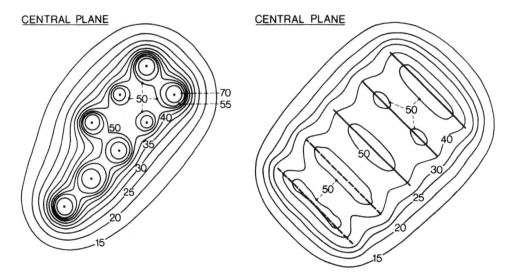

FIGURE 6.   The dose distribution of an eight-needle volume implant for breast cancer therapy. Two perpendicular sections through the central planes. The dose distribution is given in 1/10 Gy. Calculations made by the RT/PLAN (GE).

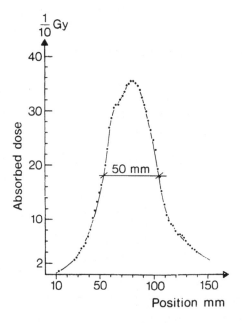

FIGURE 7.   LiF-teflon® long-rod measurement in needle nb2 (middle in upper plane) of the volume implant shown in Figure 6. Readout system TLD-20 (Therados AB).

best isodose distribution around the target volume. Only one needle at a time is exposed so the empty needles can be filled by dosimeters. As the source is switched to the next needle, this particular dosimeter is taken out. Continuous dosimeters are well suited for these measurements as they give a complete cover along the entire needle.

Figure 5 illustrates a surface implant with nine implanted needles. TL measurements are performed with LiF-teflon® long-rod dosimeters.

The computed isodose distribution of an eight-needle volume implant is depicted in Figure

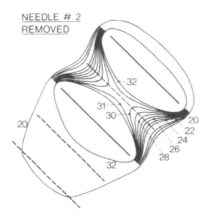

NEEDLE # 2
REMOVED

FIGURE 8. Dose distribution through the upper plane of the volume implant shown in Figure 6, with needle number 2 disabled. Dose values are given in 1/10 Gy.

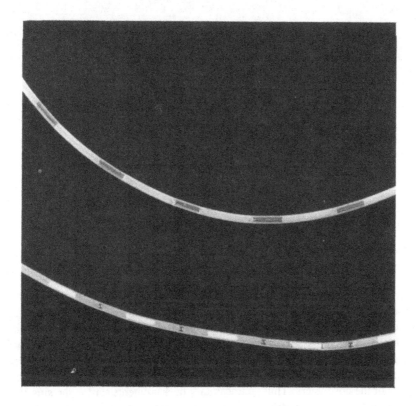

FIGURE 9. Measuring probe containing LiF-teflon® dosimeters (6 × 1 mm), and an indicator probe containing copper rods instead of dosimeters.

6. The results for needle number 2 (middle in the upper plane) are plotted in Figure 7. Note that the measured dose is valid for the dose distribution of the implant, excluding the needle in which the dosimeter was placed. The therapy planning system (RT/PLAN, GE) readily produces the corresponding dose distribution by disabling the needle number 2 (Figure 8). In this way the computed therapy plan can be compared with the results from all the needles, if necessary.

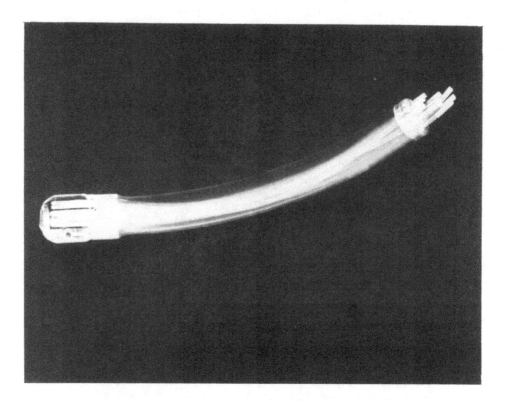

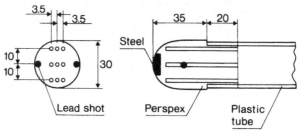

FIGURE 10.   Cylindrical rectal probe for introduction and positioning of nine TL-measuring probes. Dimension in millimeters.

1. Intracavitary Measurement Technique

For intracavitary measurements it is often necessary to sterilize the material used. As this may interfere with the thermoluminescent response of the TL dosimeters, a technique has been developed that avoids sterilizing the dosimeters themselves.[49] This technique is described in general terms and referred to below when the various applications are presented.

The insertion into the cavity must be made under radiographic control in order to determine the positions of the dosimeters. Thus, intracavitary measurements start at the simulator unit. First, a sterile plastic catheter, the outer tube, is inserted into the cavity. The catheter is proximally closed with a brass plug. The outer end of the catheter is taped to the skin when the position is satisfactory. The dosimeters are prepared in teflon® tubes called measuring probes (Figure 9). Indicator probes containing copper rods are placed in the same geometry as the measuring probes. Measuring and indicator probes can be inserted into the outer tube. The indicators are necessary because the dosimeters are not visible on radiographs. The indicator probe is inserted to the very tip of the outer tube. Frontal and lateral radiographs are taken. The indicator probe is then pulled out and the measuring probe inserted, again

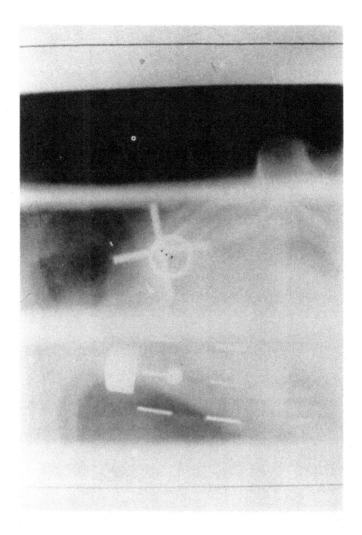

FIGURE 11. The position of the rectal probe relative to lateral skin marks and the pubis. Narrow field technique used.

to the full length of the outer tube, which means that the positions of the indicator copper rods are now taken by the TLDs. The patient is then transported to the treatment unit, without changing the position of the catheter and treatment is performed. It is of great advantage if the patient is lying down on a stretcher during the whole procedure.

The advantages with this method are that the dosimeters do not come into contact with tissue, they do not have to be sterilized, and the positions of each individual dosimeter is known from the radiographs.

2. Rectal Measurements

Rectal measurements are among the simplest intracavitary in vivo measurements that can be made. The rectum is very often considered a region of interest for treatment of the pelvis (e.g., in carcinoma of the bladder, or the cervix uteri), because overdosage in this region will cause great harm to the patient.

Intrarectal measurements have been made in a comparison of three methods of treatment of bladder cancer.[50] This is an opportunity where a computer-calculated treatment plan can easily be checked by intracavitary measurements in the organ of interest.

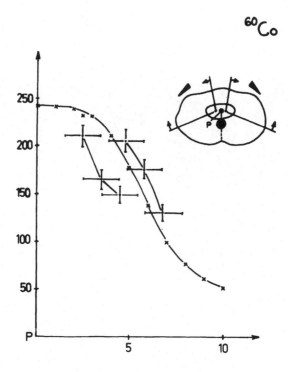

FIGURE 12. ^{60}Co arc therapy using two lateral wedged fields.
Measurements performed on two occasions. Continuous curve
corresponds to the computer-calculated dose distribution. The
three measurement points are plotted with estimated maximum
error bars.

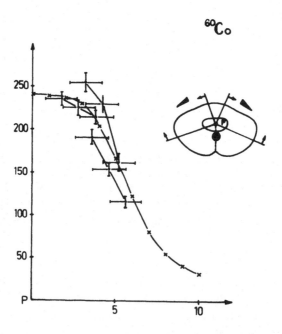

FIGURE 13.   Another patient treatment similar to the one
in Figure 12. Measurements performed on three treatment
sessions.

LA 5MV

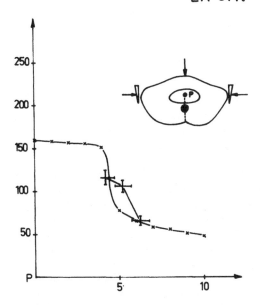

FIGURE 14.   5-MV X-ray three-field technique with lateral wedges. The absorbed dose gradient is considerably disturbed.

LA 5MV

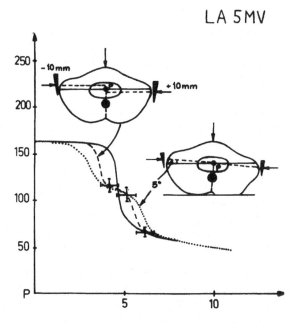

FIGURE 15.   Calculated absorbed dose distribution for two possible positional errors, ± 10-mm shift of lateral portals, and 5° tilt of pelvis. Measurements according to Figure 14.

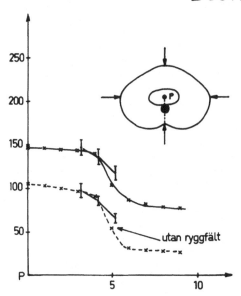

FIGURE 16.   31-MV X-ray, four-field technique. The
rectal measurements indicate a slope that is less steep
than predicted.

A rectal probe (Figure 10) is constructed in order to keep the dosimeters in a fixed geometry
inside the organ. The probe is inserted under radiographic control and turned around its axis
until the two peripheral lead shots are lined up with the horizontal beam direction.

Indicators are then inserted into the nine tubes in the probe and radiographic films are
exposed from two perpendicular angles (Figure 11). A narrow field technique is used to
avoid oblique projections and geometric magnification. Several exposures are made on the
same film, each time with a transversal shift of the position of the patient relative to the
beam. The patient is lying on a portable table-top in order to avoid changing the position
of the rectal probe during transportation from the simulator to the treatment unit.

Observations are as follows:

1.   Arc therapy with lateral-wedged ⁶⁰Co fields (Figure 12). Measurements are made on
      two occasions. The slope of the dose is in agreement with the computer-calculated
      gradient. The deviation of the dose level, however, is -25%. In another patient,
      undergoing the same treatment, it is found that the maximum deviation amounts to
      +10% (Figure 13).
2.   Three-field technique with lateral wedges using 5-MV photons (Figure 14). The cal-
      culated dose gradient through the rectum shows a steep slope which is not confirmed
      by the measurements. The maximum deviation is +30%. A careful check of the
      patient set-up shows that the skin marks, indicating the entrance points of the lateral
      portals, differ by 15 mm in height from the treatment table. This error can easily occur
      if the pelvis of the patient is not horizontally positioned during the simulation. The
      computer calculations are repeated using (1) a ±10-mm shift of the lateral fields and
      (2) a 5° angulation of the pelvis (Figure 15). In both cases the dose gradient now
      agrees with the measured values.
3.   Four-field technique using 33-MV photons (Figure 16). The measured slope is less
      steep than predicted by the calculations.

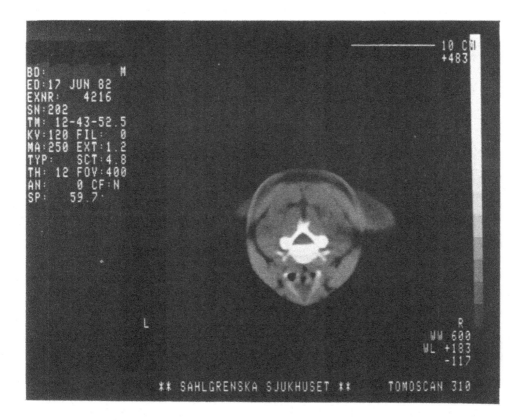

FIGURE 17.   CT scan of a Hodgkin's patient: neck section.

This study demonstrates that the day-to-day variation of the absorbed dose distribution may be considerable. The maximal difference between measured dose and computed dose reaches 30% in this study.

3. Measurements during Mantle Therapy for Hodgkin's Disease

Hodgkin's disease is a malignant lymphoma that usually arises in the lymph glands in the neck and eventually spreads throughout the lymphatic tissues of the body. The lymph nodes in the mediastinum and the lung hila, the infraclavicular fossae, both the axillae and both sides of the neck are irradiated. The lungs are not involved and must therefore be protected from irradiation by lead shields.

Hodgkin's disease is treated with two large, irregularly shaped, opposing mantle fields covering the upper half of the body, anteriorly and posteriorly. The fields extend from the mandible to the diaphragm. Figures 17, 18, and 19 show transversal radiographs, for representative parts of this region taken with a CT scanner. It is obvious that the shape, size, and homogeneity of the body change widely over this region.

The dose distribution in the body is affected by parameters such as the intensity variation across the unattenuated primary beam, the relative absorption throughout the treated volume, the irregularities of the patient contours, and the variation of the scattered radiation due to the irregular field shape. The great difference of cross section between the mediastinum and the neck is of particular concern. Measurements laterally on the neck show the highest dose values and can be 30 to 40% more than the mediastinal midplane dose. The intensity decreases at the lateral edges of the beams because of the large lung blocks that reduce the side-scattered contribution. Consequently, the greatest problem is to flatten the beam in the sagittal plane of the body.

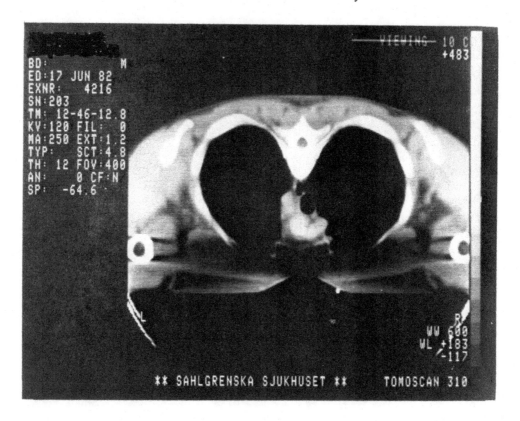

FIGURE 18.   CT scan of a Hodgkin's patient: upper thorax section.

The position of the esophagus is very closely aligned with the extension of the target volume of this disease which gives a unique opportunity for target dose measurements in vivo. Besides, when the patient is turned over in the AP-PA positions, the esophagus will closely follow the movements of the other inner organs. Thus measurements in the esophagus will give proper information about the dose distribution for the design of flattening filters taking all variability into consideration.[51]

The technique for esophagus measurements is in accordance with the general technique for intracavitary measurements described earlier. The patient is given local anasthesia in order to limit the reaction. The outer catheter is inserted either through the nose or the mouth. The treatment is given with half the prescribed dose (0.75 Gy midplane) on the anterior and the posterior fields. The treatment unit is a 5-MV linear accelerator.

Typical results of an esophagus measurement are given in Table 1. The measuring probe contains 30 dosimeters which cover a total length of about 48 mm. The principles of filter calculation are schematically illustrated in Figure 20. The absorbed dose (D), at a random point on the plotted dose curve, is the sum of the absorbed dose contribution from the anterior field (D_a) and the posterior field (D_p). The operator determines the dose level (D_0) which represents the level for which the dose distribution is flattened by the filter. The plot is placed on a curve tracer with the intersection point between the central ray and the zero level (D_0) at the origin of the tracer table. The curve is traced and the data stored in the computer memory (PDP8/E). The filter is computed in a matrix of 42 elements (N), each 5-mm wide.

A ready-made filter is depicted in Figure 21. It is constructed of aluminum slabs stuck together by tape.

The filter is mounted at 665 mm from the focus of the accelerator beam. When the filter

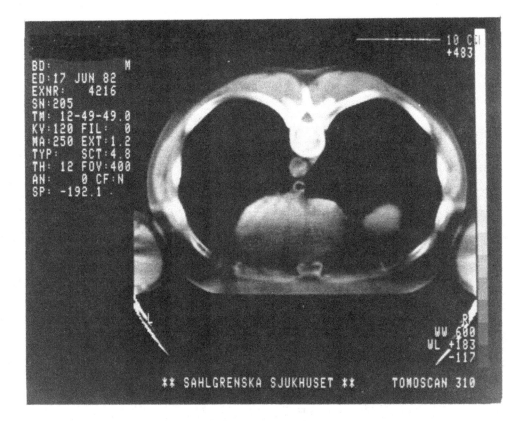

FIGURE 19.   CT scan of a Hodgkin's patient: lower thorax section.

is properly positioned in the beam and the patient aligned, the central ray passes through the center of the filter and the corresponding skin marks on the patient.

The flattening effect of the filter is checked during the following treatment session again by an esophagus measurement. Figure 22 shows the results before the filter is applied, and Figure 23 shows the result when the same patient is treated with the filter. The variation of the measured absorbed dose distribution is ±5% along the esophagus including the lateral region of the neck. Occasionally, the patient refuses to take an esophagus probe. Then, the measurements are made on the skin both at the entrance and the exit sides. The absorbed dose in the midplane can be approximately determined from these measurements and the flattening filter can then be designed in much the same way as if esophagus measurements were made.

The algorithms for the calculations are given in Appendix 5.

4. Dosimetric Studies in the Treatment of Mammary Carcinoma

Breast cancer is the most common form of female cancer. The primary tumor is locally invasive spreading along the lymph channels to the nodes in the internal mammary chain, the axillar nodes, the supraclavicular nodes, and the subclavian node. The treatment of mammary carcinoma is therefore a rather controversial subject. TL measurements have been used for in vivo studies mainly in the postoperative radiation therapy of the chest wall and the lymphatic glandulae.[52] Figures 24 and 25 illustrate the target volumes of the axillary and the supra- and infraclavicular lymph nodes and the internal mammary lymph nodes, respectively. In vivo measurements can be made in the internal mammary artery following the technique described above. First, the outer sterilized plastic tube, sealed with a gold plug at one end, is inserted into the artery. Then, the indicator probe is inserted into the

Table 1
TYPICAL PRINT-OUT FROM A DESK-
TOP CALCULATOR ON-LINE WITH
THE TLD-10 AUTOMATED
MEASURING SYSTEM

Probe no: 30.04   **Midplane dose: 1.50 Gy**

Dosimeter no.	Dose (Gy)	Midplane dose (%)
1	0.111	7.4
2	0.124	8.3
3	0.171	11.4
4	0.574	38.3
5	1.309	87.3
6	1.548	103.2
7	1.651	110.1
8	1.762	117.5
9	1.73	115.7
10	1.793	119.5
11	1.736	115.7
12	1.672	111.5
13	1.669	111.2
14	1.622	108.1
15	1.540	102.7
16	1.478	98.6
17	1.437	95.8
18	1.423	94.8
19	1.377	91.8
20	1.380	92.0
21	1.354	90.3
22	1.354	90.3
23	1.293	86.2
24	1.262	84.2
25	1.232	82.1
26	1.279	85.3
27	1.048	69.9
28	0.507	33.8
29	0.173	11.5
30	0.134	8.9

Note: The number of the measuring probe was 30-4. The tumor dose was 1.50 Gy. Measurement was made in the esophagus of a patient given upper mantle treatment without flattening filter in the beam of 5 MV X-rays.

outer tube and radiographs are taken (Figure 26). The film is placed on the patient's chest and the beam is angled 15° from the sagittal plane in order to separate the sternum and the spine. The artery runs parallel to the corpus sterni about 12 mm from its edge.

The patient is then transported to the treatment unit and therapy is given to the parasternal region with high energy electrons. The indicator probe is replaced by measuring probes during the therapy.

In vivo measurements, on the right- and left-hand sides, confirm that there is no dose reduction in this region due to reduced backscattering from the adjacent lung if the energy of the electrons is about 13 MeV. Further studies of breast cancer therapy are presented below in the sections for in vitro measurements and measurements in anatomical phantoms.

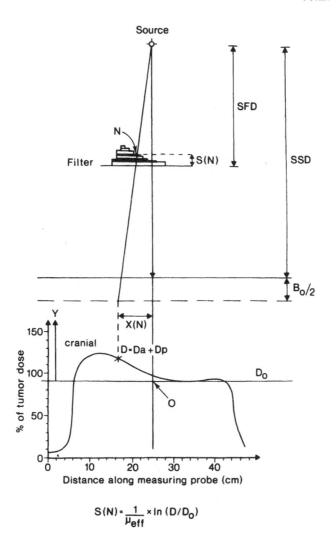

$$S(N) = \frac{1}{\mu_{eff}} \times \ln(D/D_0)$$

FIGURE 20. The principles of flattening filter calculation based on esophagus measurements. The curve is the relative absorbed dose measured in the esophagus, for the combined anterior and posterior fields. The symbols are listed in Appendix 5.

5. Dosimetric Studies in the Treatment of Thyroid Cancer

Supervoltage radiation is often given to the anaplastic variety of the thyroid carcinoma. The treatment is divided in two series of external radiation. First, high energy photons are applied, AP and PA, up to the tolerance limit of the medulla. Then, a combination of three fields are given frontally, the center beam is a high energy electron beam, with limited range, in order to save the medulla. Laterally, two photon fields are given for completion of the first treatment session. A typical therapy plan is shown in Figure 27. The range of the electrons is difficult to calculate because the trachea forms an air cavity which does not significantly attenuate the high energy electrons. Consequently, it is necessary to make esophagus measurements in order to determine the correct energy of the electrons. The method of esophagus measurement is the same as is described above, however, a dosimeter length of about 20 cm is sufficient.

Figure 28 illustrates a radiograph taken frontally over the region of interest. A measuring probe is inserted into the esophagus. This probe contains 1-mm diameter steel balls between

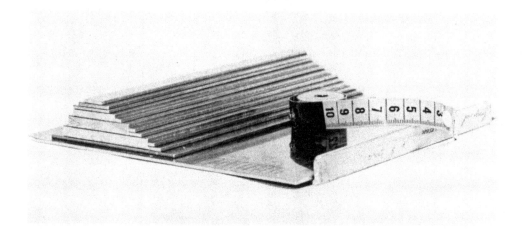

FIGURE 21.   A flattening filter composed of aluminum slabs. The scale is in centimeters.

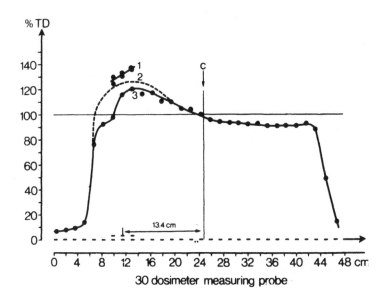

FIGURE 22.   The results of esophagus measurement without flattening filter. Measurements are also made laterally on the neck (1).

the dosimeters with double steel balls at the center. The steel balls are visible on the radiograph.

The computed isodose distribution is compared with the measurements in Figure 27. The 10% isodose line is allowed to tangent the medulla. If the measurements confirm the plan and the medulla gets below 10%, the plan is accepted.

6. Dosimetric Studies in Intracavitary Therapy

Intracavitary radiotherapy is used in the treatment of carcinoma of the cervix uteri. This is the second most common tumor in females. The most common method of treatment is radiotherapy followed by surgery. The radiotherapy is given mainly to prevent vault recurrence.

The conventional radiotherapy includes intracavitary radium sources combined with external radiotherapy to the lateral pelvic. The method is modified by accentuating the intra-

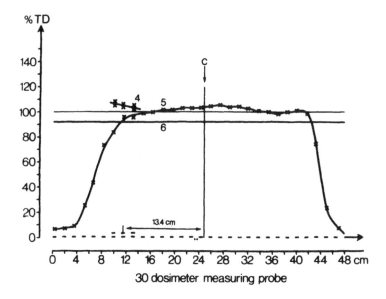

FIGURE 23. The resulting dose distribution with flattening filter for the same patient as in Figure 22.

cavitary treatment in case of predominating vaginal spread and external radiotherapy if midpelvis and parametrial involvement dominate. The adjacent walls of the rectum and the bladder must be protected in the planning.

The treatment technique is checked by in vivo intracavitary measurements in the vagina, bladder, and rectum.

Johansson et al.[49] checked the Swedish method of radium treatment by intravasal measurements in both external femoral veins using the earlier described method for intracavitary measurements. The outer tubes, consisting of sterile plastic catheters sealed with gold plugs, are inserted into both veins in cranial direction as far as the origin of the inferior vena cava. The insertion is controlled by X-ray monitoring. Figure 29 depicts a radiograph of the plastic tubes with indicator probes inserted. The radium sources are loaded and the positions of the sources and the indicators are confirmed by orthogonal radiographs. Finally, the measuring probes are inserted and the time noted. It is worth mentioning that the technique, with outer plastic catheters, has the further advantage that the measuring probes can be replaced several times during the treatment session. This improves the statistics and makes it possible to avoid overexposure of dosimeters that are close to a source. The result of the measurements indicate that the relative difference between the right and the left sides can reach 100% (Figure 30). The maximum absorbed dose is 52 Gy TBq^{-1} hr^{-1} and the minimum 20 Gy TBq^{-1} hr^{-1}, between the important positions of the principal obturator nodes and the hypogastric nodes.

The shortcoming of this treatment technique, from the dosimetric point of view, is that the positioning of the radium sources easily becomes unsymmetrical. This makes the isodose distribution also unsymmetrical, which makes it extremely difficult to combine the intracavitary and the external treatments.

Contemporary intracavitary therapy utilizes afterloaded ^{60}Co sources. An intrauterine probe is inserted and two ovoids are placed intravaginally for broadening the lateral distribution. The three sources are guided by a computer programmed for selective dose distribution. This makes it easier to apply the additional external field so that the final result becomes a homogeneous distribution of absorbed dose throughout the pelvis. In vaginal involvement the vaginal source probe is inserted into a lucite cylinder. The source can take five positions

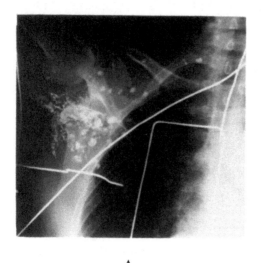

A

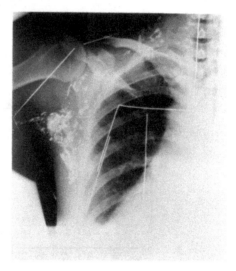

B

FIGURE 24.   The axillary and the supra- and infraclavicular nodes, after injection of Lipoidol® in a lymph vessel of the hand. Two patients (A, B) are shown. (After Ragnhult, I., Lindskoug, B., and Hultborn, A., *Acta Radiol.*, 313 (Suppl.), 135, 1972.)

in the probe. The cylinder has a track in which strings of TL dosimeters can be inserted along the axial direction. Figure 31 shows the results of an in vivo intravaginal dosimetry check. The effective length of the irradiated volume can be determined from the curve. However, as the dosimeter is attached to the lucite cylinder and not to the patient, it is mainly a check of the performance of the equipment.

D. In Vivo Measurements at Some Convenient Region

It is often too difficult to reach the region of interest by means of catheters and sometimes the patients refuse to cooperate. The measurements must then be made at some convenient region of the body adjacent to the region of interest and the data corrected to apply to the region of interest. Usually the most convenient approach is to place the dosimeters on the skin of the patient. This method has been systematized by Bagne who suggested the empirical function surface maximum ratio (SMR) described in Appendix 4.[46] SMR transforms a TL measurement on the skin to the depth d_{max}. From there the information can be transformed further to any depth by using the percentage depth dose curves. The percentage depth dose data, however, are only valid for the central-ray area. Measurements laterally in the beam must therefore be corrected to give the proper value at depth. It is worthwhile to consider the validity of the SMR function for parts of the portal outside the central-ray area. This must be checked for each treatment unit and correction factors tabulated. The sensitive layer of the skin itself is sometimes the region of interest. For relevant measurements it is necessary to know the approximate depth and thickness of the layer of interest. The average epidermal thickness is 4 mg cm^{-2} throughout the body. Distal parts of the extremities have an epidermis of 8 mg cm^{-2}, and the palms of the hands and the soles have 40 mg cm^{-2}.[53]

1. TL Measurements on the Eye

The function of the eye has been pointed out as the organ function most sensitive to radiation damage because of the risk of cataract development.[54] A special section is therefore dedicated to eye lens dose investigation.

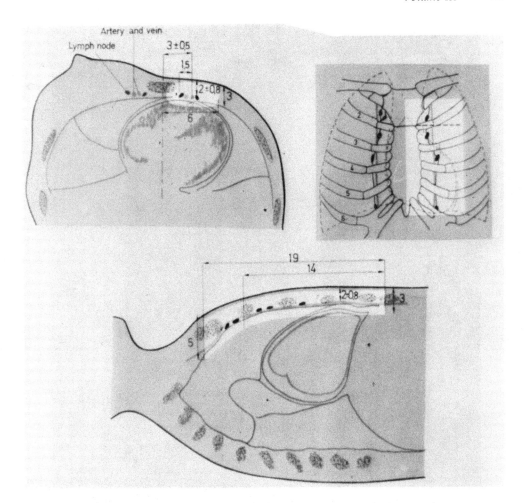

FIGURE 25. The target volume of the internal mammary lymph nodes. (After Ragnhult, I., Lindskoug, B., and Hultborn, A., *Acta Radiol.*, 313 (Suppl.), 135, 1972.)

The geometry of the eye is illustrated in Figure 32. The equator of the lens is generally at a depth of 8 mm below the eyelids. It is not possible to make direct measurements at the site of the lens, however, in many situations measurements on the eyelids give sufficient information.

Large frontal portals to the upper thorax and the neck regions produce scattered radiation to the lenses even if they are outside the primary beam. The greatest part of this radiation is scattered from blocks and collimators, while a smaller part comes from inside the irradiated body itself. For instance, in upper mantle treatments the large collimator opening exposes some of the scattering materials inside the head of the therapy unit and the large lung-shielding blocks contribute to the scattered radiation. It is therefore of great importance to protect the eye lenses in this treatment technique.

The eye lens dose is measured by means of 0.5-mm diameter × 3-mm LiF-extruded rods inserted into a polystyrene body shaped like the frontal part of the eyeball (Figure 33). The bottom of the polystyrene body is hollow and shaped to fit the eyelid. It is placed over the patient's eye during therapy and the absorbed dose is measured with and without a lead cover (Figure 34). A very simple 3-mm thick lead shield reduces the absorbed dose to the lens by about 50% when the frontal field is given. For the posterior field it is not possible to shield the lenses by simple means. The results are further checked in a Rando phantom

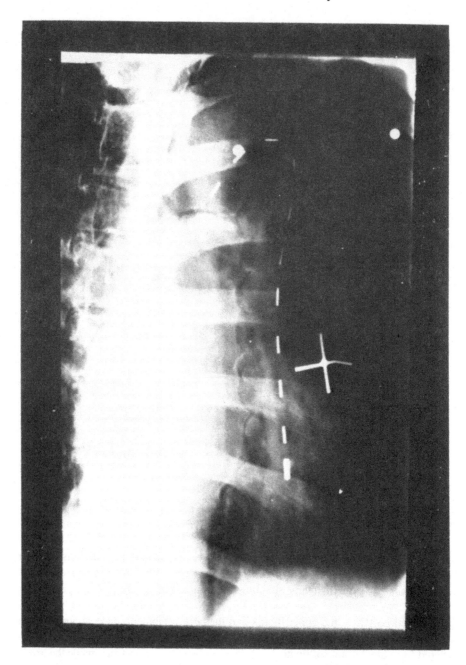

FIGURE 26. Indicators introduced into the internal mammary artery in vivo. The beam direction is angled 15° from the sagittal plane in order to separate the sternum and the spine. (After Ragnhult, I., Lindskoug, B., and Hultborn, A., *Acta Radiol.*, 313 (Suppl.), 135, 1972.)

(Figure 35) where the dosimeters are drilled into the position of the lenses. The measurements confirm the results presented above.

The eye lens dose to both eyes is determined at least during the first sessions of all head and neck treatments. The most commonly used dosimeters for this purpose are LiF-extruded chips ($3 \times 3 \times 0.9$ mm^3). The dimension relative to the eye is illustrated in Figure 32 where an uncovered chip is placed on the eyelid. If the dose is judged too high (more than

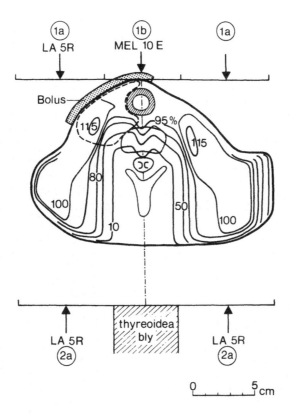

FIGURE 27. Therapy plan for treatment of cancer of the thyroid in the second series. High energy electrons are used centrally and high energy photons laterally on both sides.

2 Gy), actions are taken to block the field edge, change the beam direction, reduce the portal size, or all of the above mentioned.

Jetne developed an instrument that can be placed on the eyelid of the patient and used for finding dose-measured points on the skin from which the dose to the lenses can be deduced (Figure 36).[56] The instrument is of the "pin and arc" type in which the pin is always supposed to point at the position of the eye lenses about 8 mm below the eyelid. Each single field is checked individually. The pin is adjusted parallel to the beam axis when the arc is standing on the eyelid. The line along the pin, through the head, crosses the skin at the two measurement points. Figure 37 shows some of the typical sets of LiF-extruded rods and chips that are used in these measurements.

The effect of beam-edge blocking is studied in anatomical and regular phantoms as shown in Figures 38 and 39, respectively.

2. Measurements during Therapy of the Wilms' Tumor

Wilms' tumor is a nephroblastoma which develops from embryonic renal tissues. It is a childhood disease that is usually treated by nephrectomy followed by external radiation therapy. Bilateral involvement occurs only in about 10% of the patients. Thus it is of concern to save the healthy renal. Figure 40 shows a CT scan taken in order to display the topography and to confirm the position of the healthy renal. The corresponding dose plan, through the transversal section of the renal, is illustrated in Figure 41.

A measuring probe, containing 5 LiF-teflon® dosimeters in a teflon® tube, is placed on the treatment table below the region of interest during the first treatment session. The measured absorbed dose is presented in the figure as a percentage of the maximum dose.

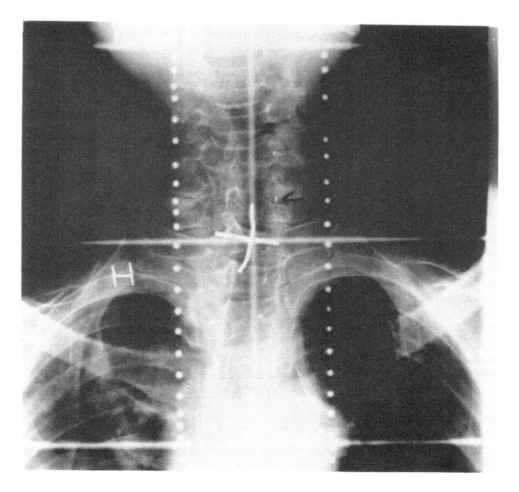

FIGURE 28.   Radiograph of the thyroid region with a measuring probe in the esophagus. The probe contains equidistant steel balls that are visible on the radiograph. The dosimeters are in the middle between the steel balls.

By extrapolating the almost-linear isodose lines at the edge of the beam, it is possible to compare the completed plan with the measured relative dose. Obviously, the practical treatment in this case corresponds well with the dose plan referring to the renal part of the dose distribution. The main part of the critical organ gets less than 20% of the maximum dose given by the frontal field. On the other hand, if it is found that the gradient of the dose along the measuring probe is less steep than planned, the portal must then be reduced or shifted laterally. Extra edge blocking may also be helpful.

Another field is applied posteriorly and directed so that the organ gets less than 5% relative dose.

3. Measurements in Connection with Whole Body Irradiation

Whole body irradiation (WBI) is used for the treatment of leukemia in connection with bone marrow transplant. This is a very unique situation in radiation therapy because the portal is much larger than the irradiated body. The field sizes used are of the order of 2 × 2 m, and the body is placed along the diagonal of the beam. All patients are treated with the same portal size and source distance. The dose distribution along the midplane of the body varies due to the influences of the scattering material brought into the field. The size and shape of the body, as well as the positions taken, interfere with the beam and thereby add to the uncertainty of the delivered dose. The prescribed dose is approximately twice the

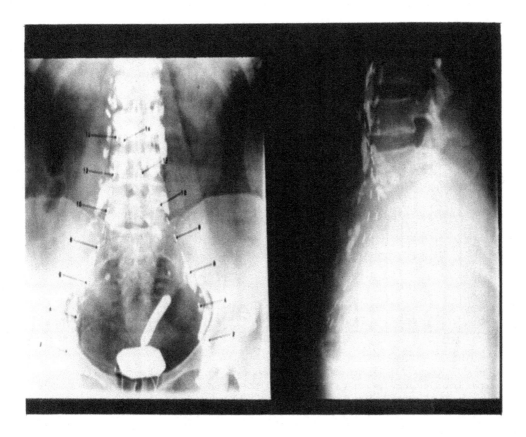

FIGURE 29. Frontal and lateral projections of intracavitary-placed radium sources. Indicator probes are inserted into the external femoral veins. (After Johansson, J. M., Lindskoug, B., and Nyström, C., *Acta Radiol. Ther. Phys. Biol.*, 8, 360, 1969.)

dose for 50% lethality had the bone marrow transplant not been given. Thus, the homogeneity and the dose level are crucial parameters for the outcome of the treatment. Fortunately, the dose is delivered at a low dose rate, usually below 0.1 Gy/minute, which gives the physicist time to make dosimetric checks in vivo.

A dose distribution variation of ±5% to the whole body is the limit for acceptance of this therapy. This goal can be reached by turning the body several times in different positions so that the exposure is given through many different angles. The patient is usually treated AP and PA with the head in the right and the left positions, then on the right and the left sides, and also with the head right and left. This adds up to eight different positions of which at least four must be checked dosimetrically.

All measurements must be made on the skin of the patient. Intracavitary or interstitial measurements are not possible because of the poor condition of the patient. Dosimeters are placed on the entrance and the exit sides of the body and proximally on the limbs and the skull. The treatment time is determined from the measurement results. Investigations of WBI of anatomical phantoms are presented in the phantom section below.

E. Measurements In Vitro

Many situations occur where in vivo measurements are difficult, but it is still necessary to get information that closely resembles the in vivo situation. In this case cadavers and fresh autopsy specimens are used. The measurements must be made within 24 hr post-mortem while the tissues are still fresh.

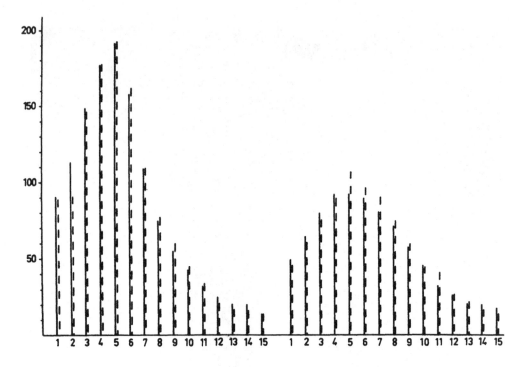

FIGURE 30.   Results of the measurements in the external femoral veins during radium therapy of carcinoma of the cervix uteri. The dashed lines correspond to a second measurement. (After Johansson, J. M., Lindskoug, B., and Nyström, C., *Acta Radiol. Ther. Phys. Biol.*, 8, 360, 1969.)

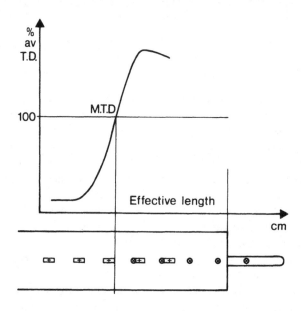

FIGURE 31.   Measurements along the intravaginal cylinder during afterloaded therapy of ca. corpus. (⊞) Dosimeters. (⊗) Source stopping points.

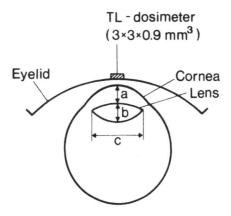

a = 3.3 ± 0.9 mm

b = 4.2 ± 1.0 mm

c = 5　±0.5 mm

FIGURE 32. The dimensions of the human eye. (After Charles, M. W. and Brown, N., *Phys. Med. Biol.*, 20, 202, 1975.)

1. Specimen Studies in Therapy for the Mammary Carcinoma

Postoperative treatment for breast cancer introduces an intriguing configuration of complex fields with various particles and energies. The target volume of mammary carcinoma treatment is described in the in vivo section above. In some patients all three lymph node regions are treated: the axillary, the supra- and infraclaviculary, and the internal mammary glands. In other patients, the axillary glands are excluded.

The supra- and infraclaviculary regions are usually treated by high energy photons, whereas the parasternal region is treated by high energy electrons. The scattering and absorption of the electrons were studied by Ragnhult et al.[52] on cadavers and behind fresh autopsy specimens of the anterior chest wall. The technique is in principle the same as the one used in vivo with outer tubes that serve as containers for the indicator and the measuring probes. The outer tubes are inserted into the mammary artery on both sides and into the brachial artery in order to cover the junction between the supra- and the infraclavicular fields and the parasternal field (Figure 42). The cadaver is brought to the simulator unit and the positions of the indicator probes are determined by orthogonal radiographs. Then the cadaver is transported to the treatment unit and complete radiation therapy is given several times. The electron energy is varied each time and the measuring probes replaced accordingly. The measurements are compared with the planned dose distribution (Figure 43). The treatment technique is then adjusted until the results demonstrate a satisfactory correspondence between the measurements and the planning.

Specimens were used for a more detailed study of the tissue equivalence of the anterior chest wall. Tissue equivalence is taken here as the equivalence to water. Irradiations are made in high energy electron fields.

Figure 44 shows a radiograph of a fresh autopsy specimen of the anterior chest wall. Two outer tubes are inserted into the mammary arteries on both sides of the specimen. Other tubes are sewed onto the back of the specimen beneath the interstices and the ribs, perpendicularly to the mammary artery. Indicator probes are used, as before, for the determination of positions. The specimen is then loaded with measuring probes and exposed to various electron energies. The measurement probes are replaced after each exposure.

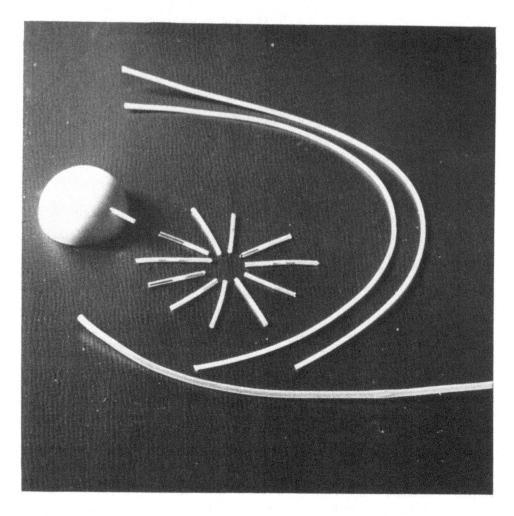

FIGURE 33.   A polystyrene phantom for checking of the dose to the eye lens. A TL dosimeter (3 × 0.5 mm diameter) in a teflon® tube is inserted into the approximate position of the eye lens equator. The small dosimeters are used for the eye lens measurements; the others are continuous dosimeters for intercomparison.

The standard deviation of the results increases significantly below 13-MeV electron energy (Figure 45). This indicates that the scattering of the electrons may be of concern if lower electron energies are used.

A more specific technique was developed for studies of the water equivalence of specimens of the rib bones and the cartilage from the thorax wall.[57] The method is illustrated in principle in Figure 46. A layer of paraffin wax (about 15-mm thick) is poured over the catheter tubes that are attached to a lucite frame through drilled holes. The specimens are pressed into the paraffin wax layer, above the catheters, just before the wax hardens. The thickness of the wax layer, that separates the catheters from the specimen, is less than 2 mm. The frame holding the specimen is then placed in a water tank at the depth of interest. Specimens mounted in the tray of paraffin wax are shown in Figure 47. The water phantom and the frame are depicted in Figure 48. The positions of the dosimeters relative to the specimens are determined from a radiograph (Figure 49).

The preparation is exposed to various electron energies at various depths in the water phantom. Then, the specimens are carefully removed from the paraffin tray without changing the positions of the catheters and the irradiations are repeated with new measuring probes

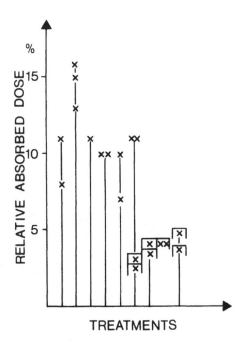

[x] 3 mm lead shield

FIGURE 34.   Results of measurements on the eye during upper mantle treatment. With (x̄) and without (x) lead shield (3 mm).

in the catheters. In this way a direct comparison is made between the water attenuation and the specimen attenuation of the electrons.

These measurements demonstrate that electron energies between 10 and 13 MeV give very good consistency between water and specimen attenuation. At energies lower than 10 MeV the attenuation in the specimens rapidly increases.

F. Measurements in Anatomical Phantoms

Existing anatomical phantoms are far from ideal. Even the best of them are designed only on the basis of radiation attenuation considerations. Mechanical properties like elasticity, mobility, and circulation, not to speak about respiration, are all neglected. An anatomical phantom is turned over like a solid body, which indeed is not the case with a living human being. However, when it comes to testing dose distributions in radiation therapy, any humanoid-shaped phantom, made by tissue-equivalent material, is far better than a square block or a parallelepipedic water tank.

Prospective and retrospective studies are usually made in anatomical phantoms. Figure 50 illustrates a Rando phantom specially constructed for continuous cylindrical dosimeters. The holes drilled through the phantoms are 2 mm in diameter and the slabs are 50-mm thick. Most commonly the phantom is used for checking the dose planning and the therapy set-up. Take the treatment of maxillary cancer as an example. Tumors of the maxillary antrum may spread to the nasal cavity, the orbit, the base of the skull, and to the tissues of the cheek. The most common treatment method is external radiation with two lateral and one frontal field, all wedged. The contralateral eye is a critical organ that must be protected. TL dosimeters are used to document the absorbed dose to the eye lens. It is necessary to specify heterogeneity corrections for the bones and the nasal cavity for the planning. At least two contours are taken from the proper levels of the phantom (Figure 51). The contours are traced into the therapy planning computer with the internal heterogeneities specified.

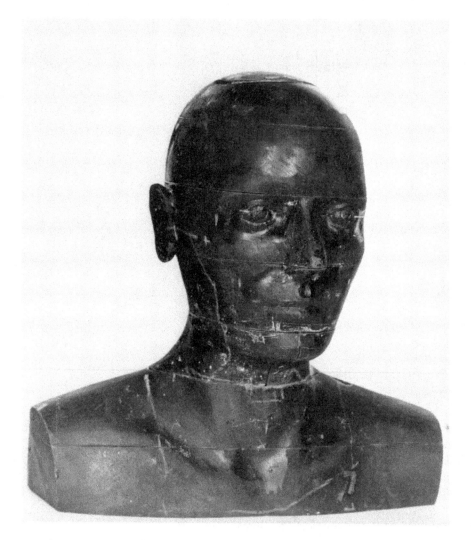

A

FIGURE 35. (A, B)   Head of the Rando phantom, with TL dosimeters drilled into the position of the eye lens equator.

The final plans are plotted and then cut along the contour lines of the skull. The plans are taped to the pertinent slabs of the phantom and dosimeters are inserted at the positions where the absorbed dose check is required (Figure 52). This technique gives the opportunity to apply the dosimeters at points right on a computer-calculated isodose line. Points of interest are behind the nasal cavity and the bone structures. TL dosimeters are also placed at the position of the eye lenses. The dosimeters are seen in Figure 52, inserted halfway into the drilled holes. The phantom is assembled and the treatment performed in much the same way as is done with an ordinary patient.

The result of a measurement may be quite surprising, particularly if the dosimeters are extended in the un-planned third dimension throughout the whole treated volume. Discrepancies of 20 or even 30%, are not unusual. Although the general aim of external therapy is to deliver a homogeneous dose in the target volume, the clinical significance of local discrepancies, which are always involved, is difficult to evaluate and is not discussed herein.

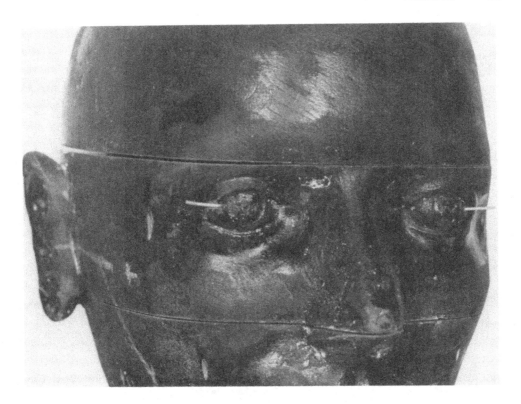

FIGURE 35B.

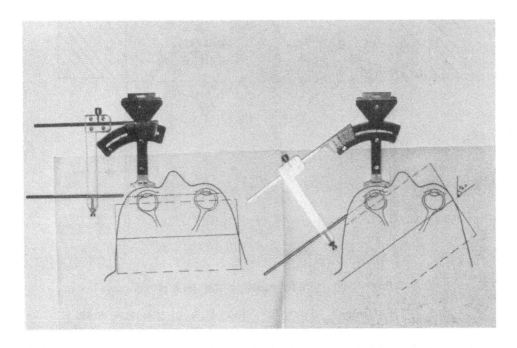

FIGURE 36. Pin-and-arc instrument for determination of eye lens relevant measurement points on the skin of the head of the patient. The pin is adjusted to be parallel to the beam axis, with the arc standing on the eyelid. (After Jetne, V., Digest of the 9th Nordic Meet. Clin. Phys., S-413 45, Roos, B., Ed., Sahlgren Hospital, Gothenburg, Sweden, 1977.)

FIGURE 37.   LiF-extruded rods, wafers, and containers. (After Jetne, V., Digest of the 9th Nordic Meet. Clin. Phys., S-413 45, Roos, B., Ed., Sahlgren Hospital, Gothenburg, Sweden, 1977.)

Epipharyngeal tumor.

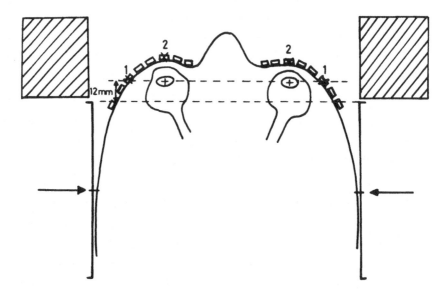

Position 1: Dose equal to the lens dose.

Position 2: Dose less than 50% of the lens dose.

FIGURE 38.   Eye lens doses measured in a skull phantom with beam-edge blocking. Position 1: dose equal to the lens dose. Position 2: dose less than 50% of the lens dose. (After Jetne, V., Digest of the 9th Nordic Meet. Clin. Phys., S-413 45, Roos, B., Eds., Sahlgren Hospital, Gothenburg, Sweden, 1977.)

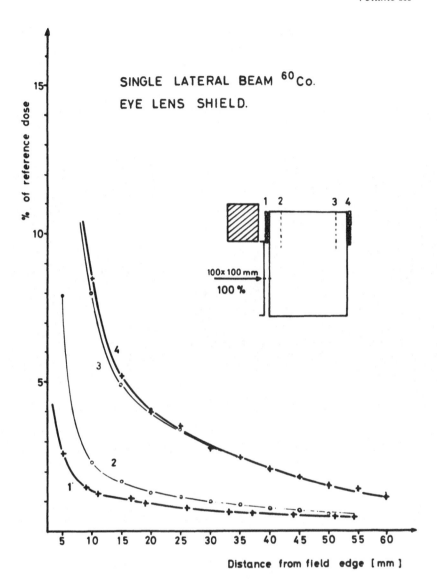

FIGURE 39. Absorbed dose measured in a square phantom, with beam-edge blocking. (After Jetne, V., Digest of the 9th Nordic Meet. Clin. Phys., S-413 45, Roos, B., Ed., Sahlgren Hospital, Gothenburg, Sweden, 1977.)

1. Phantom Studies in Whole Body Irradiation

Whole body irradiation was discussed in the section for measurements in convenient regions. As it is impossible to perform intracavitary measurements in these patients, one is forced to use total body anatomical phantoms of different sizes.

Lam et al.[58,59] investigated the dose distribution in WBI using three different sizes of anatomical phantoms: adult, adolescent, and child. Measuring probes containing LiF-teflon® dosimeters are inserted into 2-mm holes in the anatomical phantoms.

Figure 53 shows the absorbed dose distribution across the diagonal of the ^{60}Co beam without scattering material except for a concrete wall about 1 m behind the plane of measurement. The dose distributions in the midplanes along an adult and a child phantom are shown in Figure 54.

Aget et al.[60] present TL measurements in a Rando phantom compared with computer

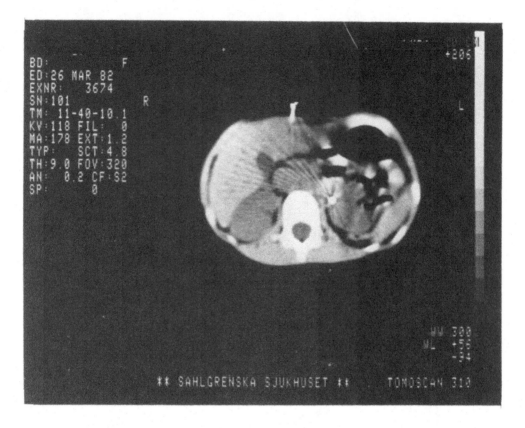

FIGURE 40.   CT scan of renal section of a 9-year-old child, for planning of Wilms' tumor therapy. The star artifact is due to clips in the operated region.

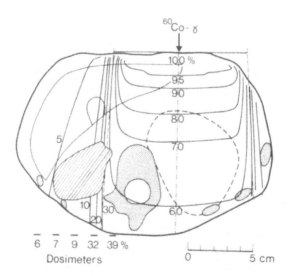

FIGURE 41.   A dose plan for treatment of Wilms' tumor. Positions of TL dosimeters and the relative results of measurement are shown.

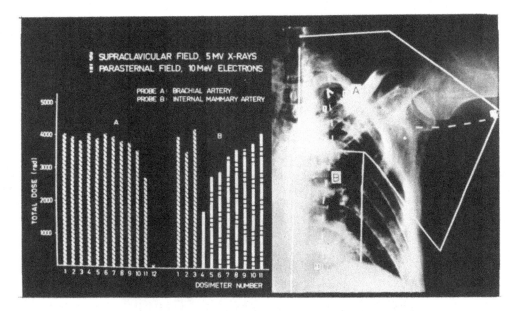

FIGURE 42.   Absorbed dose measured in a cadaver. (A) Supraclavicular field treated with 5-MV X-rays. (B) Parasternal field treated with 13-MeV electrons. Indicator probe A is inserted into the brachial artery, and probe B is in the internal mammary artery. (After Ragnhult, I., Lindskoug, B., and Hultborn, A., *Acta Radiol.*, 313 (Suppl.), 135, 1972.)

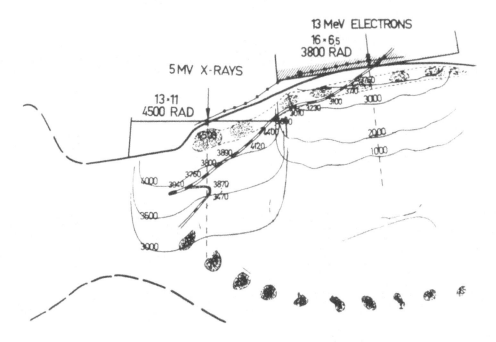

FIGURE 43.   The same case as shown in Figure 42. The measurement results are compared with the computer-calculated isodose distribution. (After Ragnhult, I., Lindskoug, B., and Hultborn, A., *Acta Radiol.*, 313 (Suppl.), 135, 1972.)

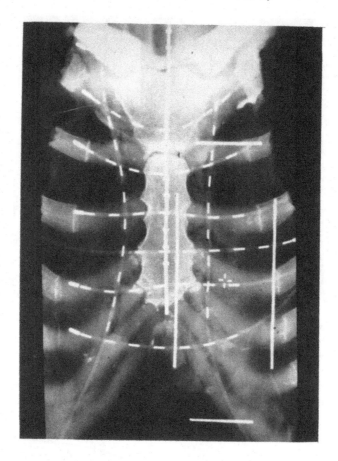

FIGURE 44.   Fresh autopsy specimen of the anterior chest wall. Indicator probes are inserted into the mammary arteries and other probes are sewed onto the back of the specimen. (After Ragnhult, I., Lindskoug, B., and Hultborn, A., *Acta Radiol.*, 313 (Suppl.), 135, 1972.)

calculations. They find a maximum variation of ±7% in a 25-MV photon beam. The measurements confirm the calculations.

2. Breast Phantom Measurements

Dosimetric studies in breast cancer therapy have been described in the sections for in vivo and in vitro measurements.

Contemporary treatment techniques strive to save the breast with a good or acceptable cosmetic result. The intact breast is usually given tangential irradiation.

It is often observed that erythema may occur in skin folds under the breasts. Questions are therefore often raised about the superficial dose distribution in the tangential fields. This was investigated by Ssengabi, who used stacks of $Li_2B_4O_7$:Mn-teflon® dosimeters with a diameter of 7 mm and 0.13-mm thickness, in a Mix-D breast phantom.[61] The diameter of the phantom is 10 cm. A selection of the results are presented in Figures 55 and 56 for different angles of incidence. These data should be compared with the patient dose measurements shown in Figure 57. The results demonstrate the shortcomings of computer planning in predicting the absorbed dose distribution superficially in an irregularly shaped body like the breast.

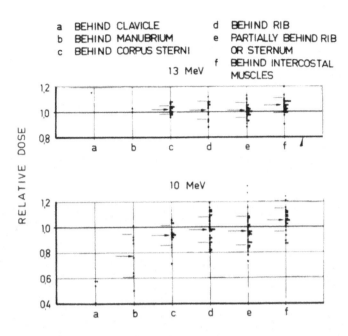

FIGURE 45. Relative dose measured behind various tissue structures. (a) Behind the claviculae, (b) Behind the manubrium. (c) Behind the corpus sterni. (d) Behind the rib bones, (3) Partly behind the ribs and the sternum. (f) Behind intercostal muscle tissue. (After Ragnhult, I., Lindskoug, B., and Hultborn, A., *Acta Radiol.*, 313 (Suppl.), 135, 1972.)

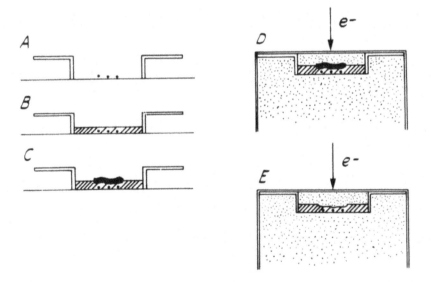

FIGURE 46. Schematic presentation of the experimental method used for testing the water equivalence of specimen of the anterior chest wall. (A) The plastic tubes and the perspex frame are placed horizontally on a level surface. (B) Molten paraffin wax is poured over the tubes. (C) The specimens are pressed down over the tubes just before the wax hardens. (D) The perspex frame with the specimen is inserted into a water tank and irradiated. (E) The specimens are removed from the wax tray. The irradiation is repeated with fresh dosimeters inserted into the plastic tubes. (After Lindskoug, B. and Hultborn, A., *Acta Radiol. Ther. Phys. Biol.*, 15, 97, 1976.)

FIGURE 47.   Fresh autopsy specimen of rib bone, rib cartilage, and sternum, mounted in paraffin wax. The plastic tubes emerging from the frame are used as containers for the measuring probes.

G. Special Phantom Measurements

Dosimetric problems can, in some situations, be satisfactorily investigated with special phantoms.

1. Build-Up and Build-Down Measurements

Secondary particles are produced at the interaction sites when ionizing radiation penetrates a body. Their direction of motion is mainly the same as the direction of the beam. The build-up region, generated when the beam penetrates into matter, extends to a depth equal to the average range of the secondary particles in the material. A similar situation occurs at the exit side because backscatter is reduced as the exit surface is approached. This phenomenon may be called "build-down".

Figure 58 illustrates schematically a phantom arrangement for the study of build-up and

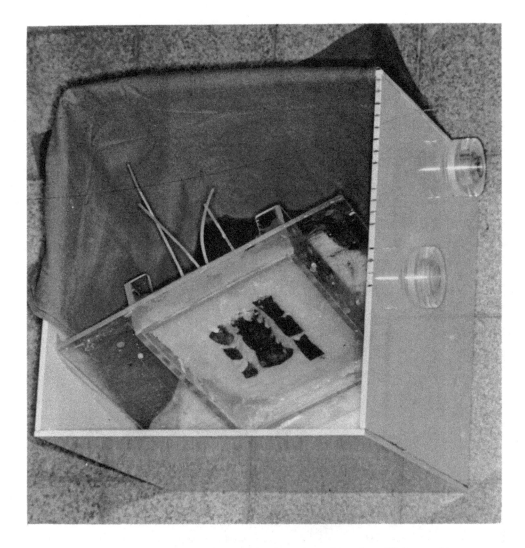

FIGURE 48. Perspex frame with specimen and catheters inserted into the unfilled water phantom.

build-down in air cavities of variable sizes. Nilsson and Schnell[62] used 10-μm LiF-teflon®
dosimeters for the study of these phenomena in a polystyrene phantom. The dosimeters are
cut from a 13-mm diameter LiF-teflon® rod using a microtome. The average thickness
of the dosimeters is 1.6×10^{-2} kg m^{-2} with a standard deviation of 0.2×10^{-2} kg m^{-2}. The
results for three photon energies (1.25, 6, and 42 MV) are shown in Figure 59.

Bertilsson made similar measurements with LiF-teflon® dosimeters in a teflon® phantom.[63]
By using only teflon® materials, this technique approaches the ideal with respect to cavity
effects since the phosphor grains, distributed in the teflon® dosimeters, introduce only a
minor cavity problem.

Build-up and build-down measurements for various-size air cavities are shown in Figures
60 and 61, respectively.

A depth dose curve for a 35-MeV electron beam is depicted in Figure 62. The influence
of air cavities of various dimensions was investigated. Even for high energy electrons there
is a remaining dose reduction after the interface build-up region compared to the reference
curve, but the depth dose is restored at the tail.

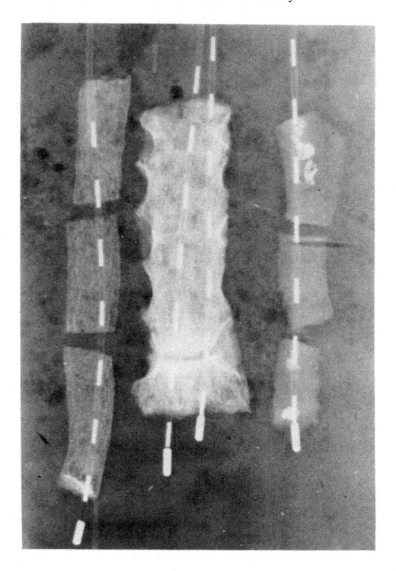

FIGURE 49.   Radiograph of the specimen tray. Indicator probes are inserted into the catheters.

2. Beta Particle Measurements

Beta particles have a continuous energy spectrum and a very short range in tissue. They introduce both an LET problem and a cavity problem to the dosimetry worker.

Kastner et al.[64] studied these problems in an investigation of the dose distribution around a needle containing a $^{90}Sr-^{90}Y$ source. They used LiF-teflon® dosimeters with 30% loading cut in dimensions of 0.1×1 mm diameter. It is possible to use these small dosimeters to determine the dose distribution around the needle.

Greitz and Rudén[65] studied the internal distribution of beta-ray dose. The self-shielding effect is measured in aqueous solutions of various beta emitters. The dosimeters are completely immersed in the solution at the center of the container and with the dosimeters closely pressed against the wall. The result is demonstrated in Figure 63.

3. Phantom Measurements in Interstitial Afterloading

The principles of computerized interstitial afterloaded therapy were described in the section for in vivo measurements. Such a system is used by Schulz for the treatment of superficial

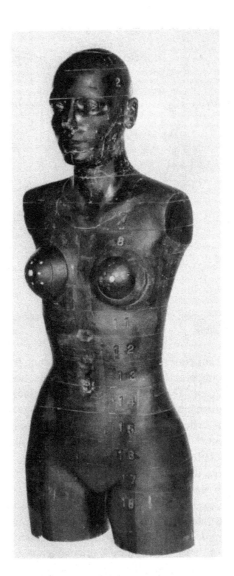

FIGURE 50.   A specially designed Rando
phantom with 50-mm thick slabs.

tumors.[66] LiF-teflon® rods in teflon® tubes (TLD-10, Therados) are used for checking the
dose distribution around the needles.[67] The polystyrene phantom depicted in Figure 64 is
used for simulating the patient. The phantom is loaded with measuring probes that fit into
tracks in some of the plates (marked TLD). The implanted needles are placed in similar
tracks in other plates (Nadel). Then the source container is connected via a tube and the
irradiation performed.

Results for a three-needle implant are shown in Figure 65. The doses measured are
presented in four planes with computer-calculated isodose lines. They agree very well with
the measured values which indicates that the calculation of the exposure times at the source
stopping points in the needles are correct and that the afterloading equipment is working
properly.

Advanced computerized interstitial afterloading gives very complicated dose distributions
that are difficult to check with conventional dosimetry. Measurements at a single, or a few

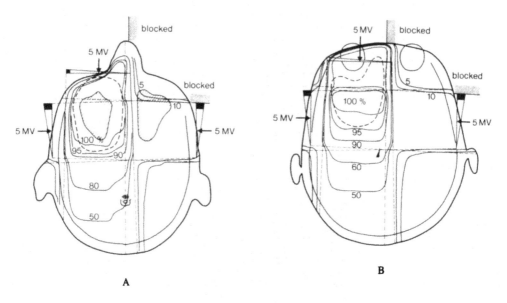

FIGURE 51.   Therapy plans for treatment of the maxillary antrum in the Rando phantom skull. (A) Through the orbita. (B) Through the nasal cavities.

points, are not sufficient because one, or several, source stopping points might have exposure time errors of considerable importance for the adjacent tissues, not observable, however, at distant positions in the implant. This arises from the steep gradient of dose around point sources. Adequate measurements were made by Lindskoug using continuous dosimeters.[68,69] The principles of continuous cylindrical dosimeter readout by linear motion through a heated oven are described elsewhere.[45] The basic advantage of these dosimeters is that the absorbed dose can be measured at a continuous series of points, i.e., along a line, independent of the curvature of this line in three-dimensional space.

The implant is made in a water tank by fixing the needles in polystyrene frames (Figure 66). Tube wall dosimeters are inserted between the needle planes.[70]

A simple example is illustrated in Figure 67. This is a four-needle quadratic implant with a needle separation of 30 mm. The intention is to check the activity of the source and the performance of the irradiator. Two U-bent tube wall dosimeters are placed in the mid-plane between the needles where the absorbed dose distribution goes through a minimum (Figure 68). This reduces the demands for high precision positioning of the dosimeters and it also eliminates the problems involved with measurements in beams of steep dose gradients as is the case around a single source.

The results of the measurement for one of the dosimeters are shown in Figure 69. The same exposure time is used at all source stopping points along the full length of the needles (100 mm). The readout integration interval is 1 mm. The averages of the symmetrical peaks are plotted in Figure 70 and compared with computer-calculated data using the conversion factor 0.127 Gy mm^2 MBq^{-1} hr^{-1}. The results are in good correlation.

4. Delivery Checks of Radioactive Needles and Seeds

Delivery checks of radionuclides such as strings, needles, or seeds are generally performed by means of well-type crystals or ionization chambers. These instruments, however, do not give detailed information about the activity distribution along the sources.

A more detailed check of line sources and seeds can be established by means of continuous tube wall dosimeters.[70] Seeds of ^{192}Ir are delivered in strings of nylon usually containing 12 seeds. For delivery check the nylon string is pulled through a tube wall dosimeter surrounded

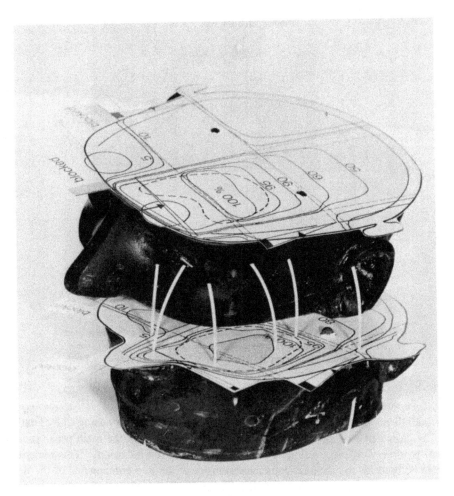

FIGURE 52. The therapy plans attached to the Rando skull section. Dosimeters are drilled into relevant positions for checking the treatment technique and the therapy planning.

by polystyrene plates. The seeds are then quickly pulled out of their lead container and into the tube wall dosimeter where they are left for about 5 min. The exact exposure time is noted. The dosimeter is read out and the result plotted (Figure 71). The diagram also shows the response of the same dosimeter in a homogeneous ^{60}Co beam. After calibration with sources of known activity, the activity of each seed can be derived.

An ^{192}Ir needle is measured in a similar way (Figure 72). It is inserted into the tube wall dosimeter for a couple of minutes and then quickly removed. The resulting readout is shown in the figure. The integration interval is 3 mm.

H. TL Measurements for Neutron Therapy

1. Introduction

The increasing use of neutron therapy stresses the need for routine clinical dosimetry in these complicated fields. Neutron dosimetry is more difficult than the dosimetry of conventional particles because the neutron field is always accompanied by gamma rays that originate in the target and from interactions with the collimators, the shielding material, as well as from the irradiated body itself. It is necessary to separate the dose distribution of the various components because they have very different relative biological effectiveness (Table 2).

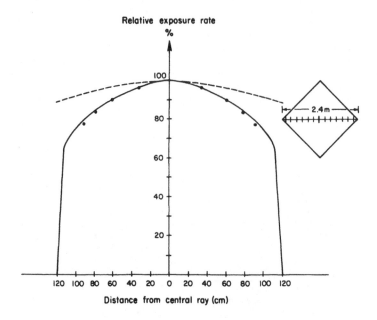

FIGURE 53.   The beam profile scanned along the diagonal of a ^{60}Co field used for WBI. The dashed curve is the expected beam variation from a point source. (After Lam, W-C., Lindskoug, B. A., Order, S. E., and Grant, D. G., *Int. J. Radiat. Oncol. Biol. Phys.*, 5, 905, 1979.)

2. Two-Component Model

Rassow[71] developed a system for routine clinical neutron dosimetry using CaF$_2$:Tm (TLD-300). The glow curve of this phosphor has five peaks and two of the main peaks (numbers 3 and 5) have very different efficiency to photon and neutron radiation.[72] This material can therefore be used for the simultaneous measurement of the two components of the neutron beam. A typical glow curve is illustrated in Figure 73. The measurements were made with a Harshaw® TL reader on-line with a HP 85 computer. The parameters of each glow peak (peak height, peak area, and the partial area between preselected limits) are mathematically computed according to a Gaussian peak-shape model.* Temme et al.[73] found that this material can be used without annealing if the integrated peaks are used as the measure of the TL response.

The total absorbed dose (D$_T$) and the gamma ray-absorbed dose contribution (D$_G$) are determined from Equations 1 and 2:

$$M_3 = a_3(D_T - D_G) + b_3 D_G \qquad (1)$$
$$M_5 = a_5(D_T - D_G) + b_5 D_G \qquad (2)$$

where M$_{3,5}$ is the measured signal for peaks 3 and 5, respectively. The a factors are the neutron sensitivities and the b factors are the gamma-ray sensitivities for the two peaks.

The calibration is made in an A-150 phantom block exposed to the d(14) + Be neutron beam of the CIRCE cyclotron.[74] The phantom block is a 30 cm cubicle. The a and b factors are evaluated by many measurements at various depths in the phantom using various field sizes. It is noticed that M/D$_T$ is a linear function of D$_G$/D$_T$:

---

*   The Gaussian peak shape is chosen to facilitate the mathematical analysis. See Volume I, Chapter 4, Section I.L for a discussion of computerized glow curve analysis using first order kinetic glow peak shapes.

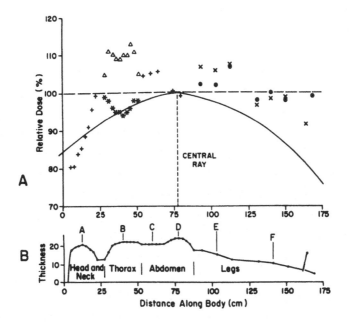

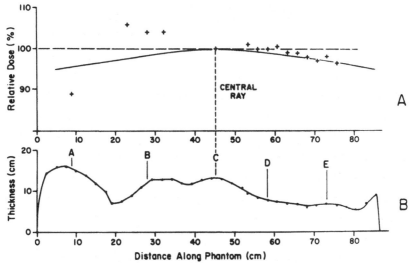

FIGURE 54. Dose distribution along an adult (top) and a child (bottom) phantom. (A). The symbol + indicates average dose in a transverse section for head and abdomen; * is the average in the sagittal plane through the midline; Δ is the average in the parasagittal plane at 7.5 cm from the midline through the right lung; • represents sections through left limbs; and x represents sections through right limbs. The continuous line represents the beam exposure profile at the midline of the phantom. (B). Phantom thickness measured through the thickest part of the section. (After Lam, W.-C., Lindskoug, B. A., Order, S. E., and Grant, D. G., *Int. J. Radiat. Oncol. Biol. Phys.*, 5, 905, 1979.)

$$M/D_T = a(1 - D_G/D_T) + b \, D_G/D_T = a + (b - a)D_G/D_T \tag{3}$$

M/D_T is then plotted as a function of D_G/D_T (Figure 74). By extrapolating to zero and reading the slope of the line, the response-coefficients can be determined. The results are presented in Table 3. This method of measurement has been tested during neutron therapy with the CIRCE.

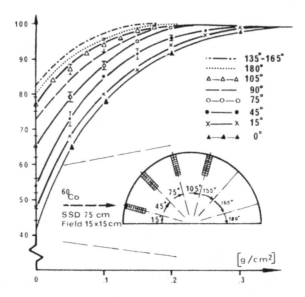

FIGURE 55.   Relative absorbed dose build-up in a breast phantom with tangential irradiation of ^{60}Co photons. The radius of the phantom is 5 cm and the field size at the entrance point of the central ray is 15 × 15 cm^2. Note the build-down effect on the exit side. (After Ssengabi, J., Studies of Photon Beam Properties in Therapeutic Applications, Ph.D. thesis, University of Stockholm, Stockholm, Sweden, 1978.)

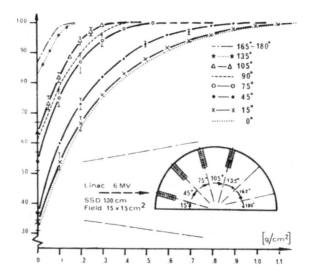

FIGURE 56.   Relative absorbed dose build-up for 6-MV linear accelerator photons. Dimensions are the same as in Figure 55. The exit dose is less dependent on the backscatter reduction for this quality than is the case in the ^{60}Co beam according to Figure 55. (After Ssengabi, J., Studies of Photon Beam Properties in Therapeutic Applications, Ph.D. thesis, University of Stockholm, Stockholm, Sweden, 1978.)

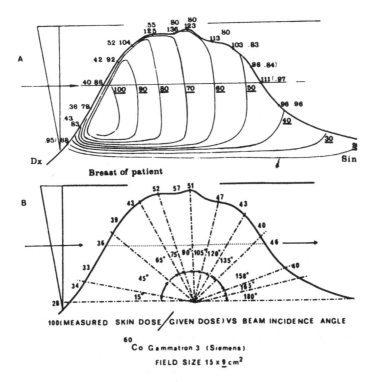

FIGURE 57. (A) Absorbed dose measured on the skin of a patient treated with a tangential ^{60}Co beam. Dose values are specified in cGy for a given dose at d_{max} of 238 cGy. Values within brackets are relative doses referring to the dose at 5-mm depth. Comparison is made with a computer-calculated isodose diagram. (B) Surface dose per 100 cGy given dose. (After Ssengabi, J., Studies of Photon Beam Properties in Therapeutic Applications, Ph.D. Thesis, University of Stockholm, Stockholm, Sweden, 1978.)

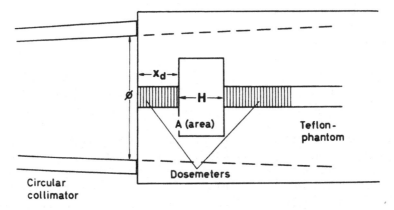

FIGURE 58. A phantom block arrangement for build-up and build-down measurements in air cavities. (After Bertilsson, G., Electron Scattering Effects on Absorbed Dose Measurements with LiF Dosimeters, Ph.D. thesis, University of Lund, Lund, Sweden, 1975.)

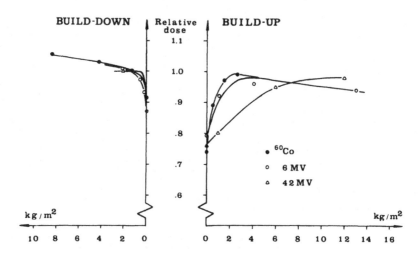

FIGURE 59.   Build-up and build-down measurements in an air cavity. Field size 40 × 40 mm², cavity depth 21 mm. (●) ⁶⁰Co gamma. (○) 6 MV. (△) 42 MV. (After Nilsson, B. and Schnell, P.-O., *Act. Radiol. Ther. Phys. Biol.*, 15, 427, 1976.)

Phantom studies are also made in order to verify the computer planning.[75] Figure 75 shows a plan for an A-150 body-shaped phantom exposed to neutron arc therapy. The isodose lines present the total dose (D_T, Gy). Figure 76 illustrates the gamma-ray component in fractions of the total dose, and Figure 77 shows the gamma-ray component only. The underlined values refer to the isodose lines and the dots are measurement points.

Other phosphors of interest for neutron dosimetry are ⁶LiF (TLD-600) and ⁷LiF (TLD-700). TLD-600 is very sensitive to thermal neutrons due to the 942 barn cross section of the ⁶Li(n,α)τ reaction.

On the other hand, TLD-700 has almost negligible sensitivity to neutrons in the low and intermediate energy range. The sensitivity to photon radiation, however, is very similar for both phosphors. Consequently, a combined measurement with both materials can be used for neutron-gamma ray separation.

3. Three-Component Model

A three-component model was used by Lindskoug for neutron beam dosimetry with the three phosphors TLD-100, -600, and -700. The isotopic content of the materials is shown in Table 4. The measurements are made with the powders enclosed by teflon® tubes with an inner diameter of 1 mm and a wall thickness of 0.25 mm. The length of the sensitive material is about 80 mm. The exposures are made in different materials such as water and polystyrene containing hydrogen, or air and teflon® with 25% carbon admixture as non-hydrogeneous materials. All phantoms have the dimensions 14 × 14 × 8.5 cm³. The water is contained in a tank with 2-mm polystyrene walls.

The dosimeters are read out in the TLD-20 system (Therados AB) for continuous dosimeters. The response is expressed per millimeter, i.e., the integrated signal along the dosimeter string divided by the length of the dosimeter. The response of the individual dosimeters is standardized by ⁶⁰Co reference irradiation according to the method described in Appendix 3. The average sensitivity to ⁶⁰Co gamma rays for the three materials is shown in Table 5.

The neutron response is standardized by the ratio,

$$S = \frac{R_{neutrons}}{R_{Co}} D \qquad (4)$$

Relative percentage depth dose

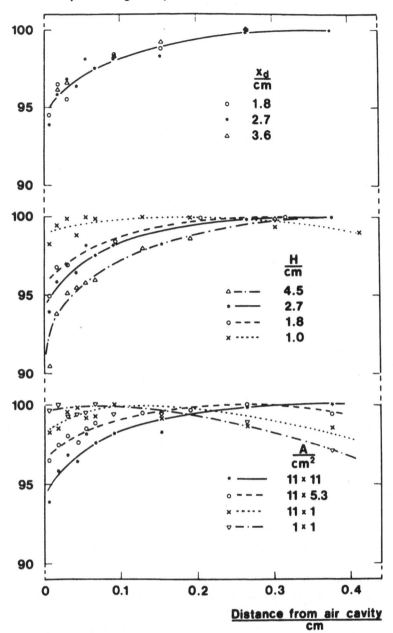

FIGURE 60. Absorbed dose build-up behind air cavities in a teflon® phantom irradiated with 35-MeV electrons. Field diameter is 8 cm. Top: air layer of height (H) 2.7 cm located at different depths (x_d). Middle: air layers of different heights at a depth of 2.7 cm. Bottom: air cavities with different cross-sectional area, height 2.7 cm at a depth of 2.7 cm. The depth doses are given relative to the maximum absorbed dose in front of the cavities. (After Bertilsson, G., Electron Scattering Effects on Absorbed Dose Measurements with LiF Dosimeters, Ph.D. thesis, University of Lund, Lund, Sweden, 1975.)

Relative percentage depth dose

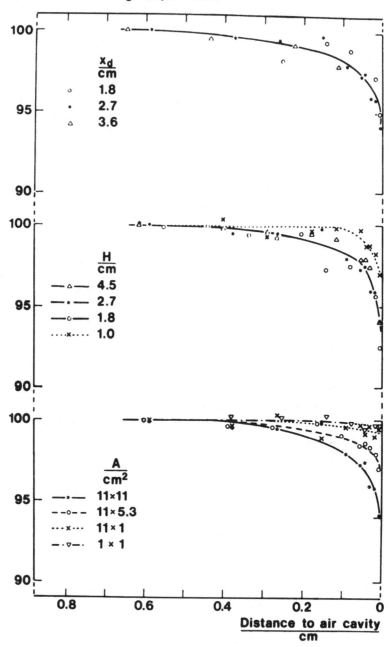

FIGURE 61. Absorbed dose build-down in front of air cavities in a teflon®
phantom irradiated with 35-MeV electrons. The field diameter is 8 cm. Top:
air layer of height 2.7 cm located at different depths x_d. Middle: air layers of
different heights at a depth of 2.7 cm. Bottom: air cavities with different cross-
sectional area, height 2.7 cm at a depth of 2.7 cm. The depth doses are given
relative to the maximum absorbed dose in front of the cavities. (After Bertilsson, Electron Scattering Effects on Absorbed Dose Measurements with LiF
Dosimeters, Ph.D. thesis, University of Lund, Lund, Sweden, 1975.)

Percentage depth dose

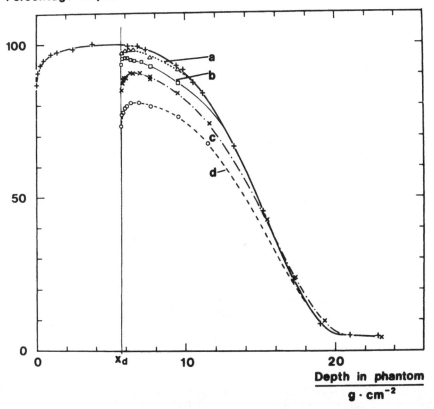

FIGURE 62. Percentage depth dose curves for 35-MeV electrons. Field diameter is 8 cm. The (+) marked curve is taken in a homogeneous medium. The outer curves are behind four different air layers at a depth of 2.7 cm with the height of the air layer equal to: (a) 0.9 cm, (b) 1.8 cm, (c) 2.7 cm, and (d) 4.5 cm. Note that the depth is given in g cm^{-2}, which reduces the height of the air cavities to approximately zero. (After Bertilsson, G., Electron Scattering Effects on Absorbed Dose Measurements with LiF Dosimeters Ph.D. thesis, University of Lund, Lund, Sweden, 1975.)

where R is the response and D is the ^{60}Co calibration dose. S is referred to as the standardized response in units of "^{60}Co equivalent Gy".

The mathematical model is composed of the following components:

Fast-neutron contribution: $k_n^f\, c_n(E_f)\, \phi_f$

Slow-neutron contribution: $k_n^s\, c_n(E_s)\, \phi_s\, sf_n(E_s)$

Gamma-ray contribution: D_γ

where n = 1, 6, and 7 for TLD-100, -600, and -700, respectively, $k_n^{f,s}$ = the ratio of the sensitivity to kerma of neutrons and the sensitivity to ^{60}Co gamma rays for fast (f) and slow (s) neutrons in the material n; $c_n(E_f)$ = the kerma factor for dosimeter material n at the neutron energy E_s; and D_x = the gamma ray-absorbed dose contribution.

The standardized response, defined above, is then given by the expression:

$$S_n = k_n^f\, c_n(E_f)\, \phi_f + k_n^s\, c_n(E_s)\, sf_n(E_s)\, \phi_s + D_\gamma \qquad (5)$$

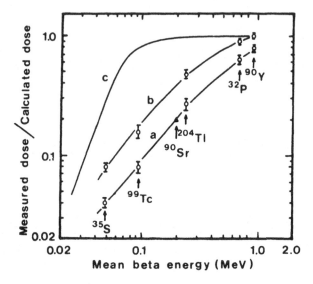

FIGURE 63.   Self-shielding effect in TL dosimeters irradiated by various beta emitters in aqueous solutions. Relative response of TL in LiF:Mg,Ti as a function of the mean beta energy: (a) with powder (0.15-mm average grain diameter) exposed at the surface of a gelatine mold;[65] (b) with 1-mm diameter LiF-teflon® rods, at the surface of the container; and (c) submerged in the solution of the nuclide.[64]

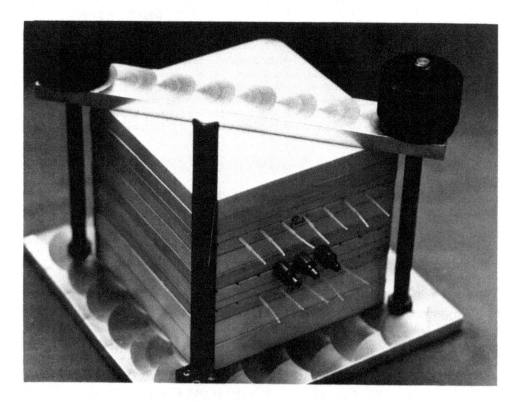

FIGURE 64.   Polystyrene phantom used for absorbed dose measurements in interstitial computerized afterloading of ¹⁹²Ir source. (After Quast, U. and Borman, U., Experimentelle und Klinische Leukämie und Tumorforschung an der Universität Essen-Gesamthochschule, Medizinische Physik 80, Heidelberg, 2, 495, 1980.)

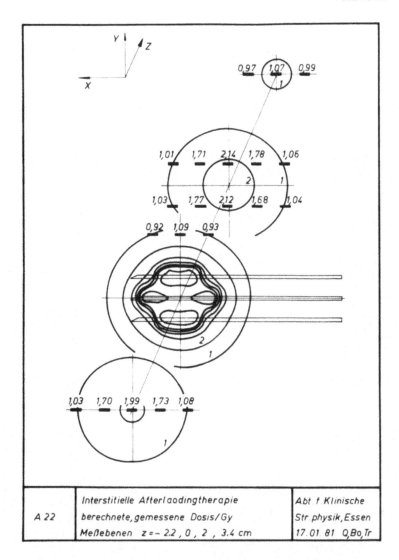

FIGURE 65. Computed isodose lines compared with measured absorbed dose for a three-needle phantom implant. (After Quast, U. and Borman, U., Experimentelle und Klinische Leukämie und Tumorforschung an der Universität Essen-Gesam thochschule, Medizinische Physik 80, Heidelberg, 2, 495, 1980.)

The fast neutron parameters are determined by calibrating the dosimeters in known beams of 14 MeV D-T neutrons, 8- and 20-MeV cyclotron neutrons (D-Be) using the hydrogenous phantom materials (water and polystyrene), and the nonhydrogenous materials (air and teflon® + 25% C). The system of equations can be solved by making certain assumptions and approximations for the different materials.

Typical standardized depth response curves are shown in Figure 78. The corresponding calibration factors are presented in Figure 79. The slow neutron parameters are determined by calibration in a fission neutron beam. Typical standardized depth response curves are shown in Figure 80. The corresponding calibration results are presented in Table 6.

a. Self-Shielding

Horowitz et al.[76] pointed out that self-shielding of the dosimeters must be considered in slow neutron beams. Values of the self-shielding factors in various geometries are tabulated

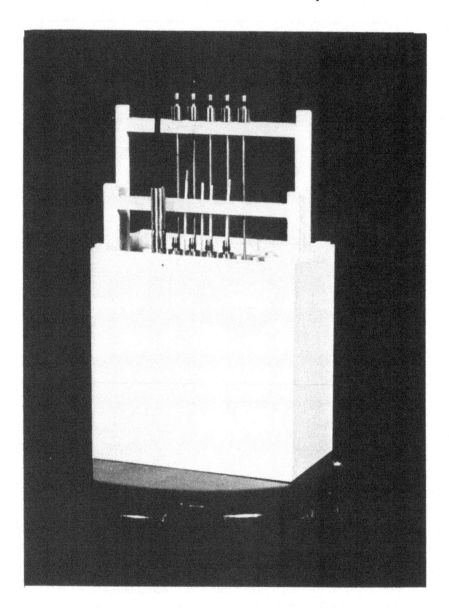

FIGURE 66.   Water phantom used for dosimetry with computerized interstitial afterloading. Implanted needles are shown. Continuous tube wall dosimeters in a polystyrene frame are inserted between the needle planes. Dimensions of phantom are $14 \times 14 \times 8$ cm^3.

in Volume II, Chapter 2, Tables 9 and 10 and can also be theoretically determined using ANISN.[77] The neutron fluence rate is calculated in a cylindrical water tank (14 cm in diameter) with and without dosimeter materials at the center of the phantom. The self-shielding factors are then calculated according to

$$sf = \phi_d/\phi_w \tag{6}$$

where ϕ_d is the calculated neutron fluence with the detector in the tank and ϕ_w is the calculated fluence without the detector. The ratio (sf) for ⁶LiF dosimeters is plotted as a function of the phantom radius in Figure 81 and the corresponding results for TLD-100 are shown in Figure 82. ⁷LiF does not show any significant self-shielding for any energy interval.

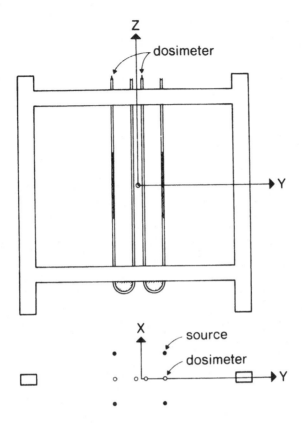

FIGURE 67.   Illustration of a measurement set-up. Top: frame holding two U-bent continuous dosimeters for insertion into the water phantom. Bottom: transversal section showing the relative positions of the dosimeters and four needles. Needle separation is 30 mm.

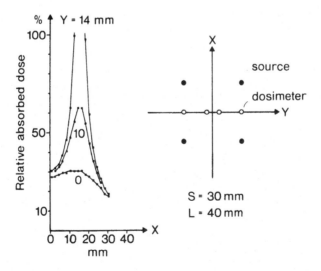

FIGURE 68.   Computed relative dose distribution in the midplane and at Y = 10 and 14 mm between the 4 sources.

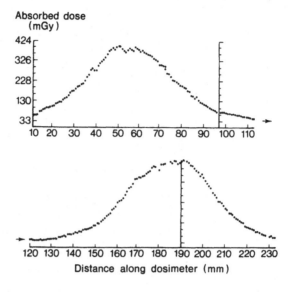

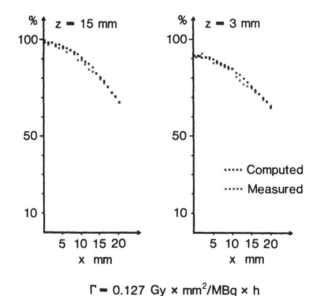

FIGURE 69.   Absorbed dose distribution measured by one of
the dosimeters shown in Figure 67. This is a computer printout
with automatic scaling.

FIGURE 70.   Sum average of the two symmetrical peaks shown in
Figure 69. The calculated distribution is in good agreement with measured values when the conversion factor 0.127 Gy mm² MBq⁻¹ hr⁻¹ is
used.

It is obvious from the diagrams that self-shielding in TLD-100 and TLD-600 is of significant importance for energies below about 1 keV. The average self-shielding factors for
the materials at some pertinent energy intervals are presented in Table 7.

b. Results

The parameters deduced from the calibration in the known neutron beams are inserted
into Equations 5. The result for 14 MeV neutrons becomes

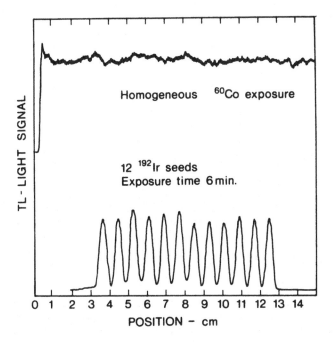

FIGURE 71. TL light emission of a tube wall dosimeter after internal exposure by 12 seeds of ^{192}Ir during 5 min. The top curve shows the corresponding result after homogeneous ^{60}Co exposure with known dose.

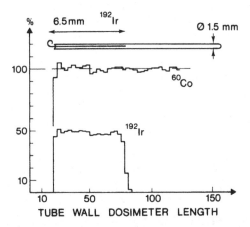

FIGURE 72. TL light emission of a tube wall dosimeter after internal exposure by an ^{192}Ir needle. The top curve shows the corresponding readout after homogeneous exposure to ^{60}Co gamma rays.

$$S_1 = 0.48 \; c_1(E_f) \; \phi_f + 0.15 \; c_1(E_s) \; sf_1(E_s) \; \phi_s + D_\gamma \tag{7}$$

$$S_6 = 0.79 \; c_6(E_f) \; \phi_f + 0.16 \; c_6(E_s) \; sf_6(E_s) \; \phi_s + D_\gamma \tag{8}$$

$$S_7 = 0.38 \; c_7(E_f) + D_\gamma \tag{9}$$

Table 2
RELATIVE BIOLOGICAL EFFECT (RBE)
FOR VARIOUS PARTICLES AND ENERGIES

Particle/quality	Energy/Eff-energy	RBE
X-ray 55 kVp/0.5 Al	17 keV	1.31
X-ray 85 kVp/1 Al	25 keV	1.29
X-ray 200 kVp/0.45 Cu	43 keV	1.20
X-ray 250 kVp/0.9 Cu	58 keV	1.18
X-ray 250 kVp/2.7 Cu	98 keV	1.16
^{60}Co gamma	1.25 MV	1.00
electrons	6—50 MeV	1.00
X-ray	2 MV and above	1.00
Protons	Excluding peak[a]	1.00
Neutrons	Fast	3

[a]   In the Bragg peak, RBE increases to about 3, but this
is only over a few microns in tissue at the end of the
particle track.

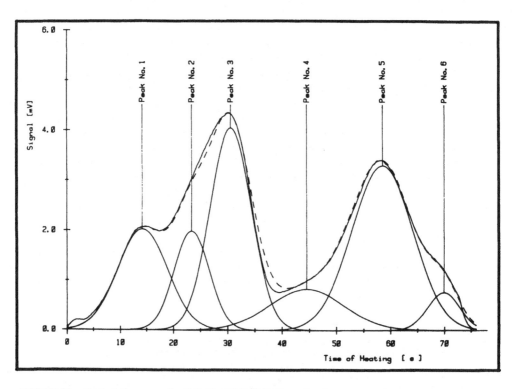

FIGURE 73.   Typical glow curve for CaF$_2$:Tm (TLD-300) exposed to fast neutrons. Each single peak is mathematically computed according to a Gaussian peak shape model. (After Rassow, J., Advisory Group Meeting on Advances in Dosimetry for Fast Neutrons and Heavy Charged Particles for Therapy Applications, International Atomic Energy Agency, Vienna, June 14 to 18, 1982.)

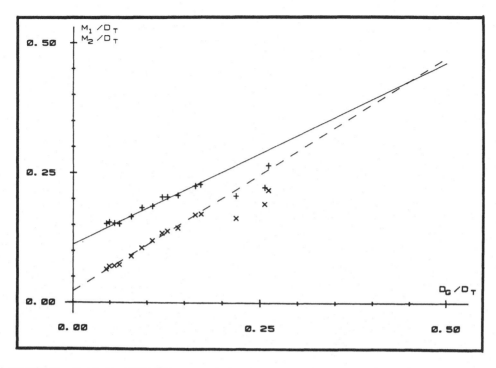

FIGURE 74.   Standardized TLD-300 response vs. the absorbed dose fraction D_G/D_T, where D_T is the total neutron + gamma-ray dose and D_G is the gamma-ray component only. (After Rassow, J., Advisory Group Meeting on Advances in Dosimetry for Fast Neutrons and Heavy Charged Particles for Therapy Applications, International Atomic Energy Agency, Vienna, June 14 to 18, 1982.)

Table 3
RESPONSE
COEFFICIENTS FOR
D(14) + BE
NEUTRONS[71]

$$a_1 = 0.022 \qquad b_1 = 0.920$$
$$a_2 = 0.113 \qquad b_2 = 0.810$$

Using the following data

$$c_1(E_f) = c_7(E_f) = 0.45 \ 10^{-8} \ \text{cGy cm}^2$$
$$c_6(E_f) = 0.67 \ 10^{-6} \qquad \text{cGy cm}^2$$
$$c_1(E_s) = 0.15 \ 10^{-6} \qquad \text{cGy cm}^2$$
$$c_6(E_s) = 0.19 \ 10^{-5} \qquad \text{cGy cm}^2$$
$$sf_1(E_s) = 0.78 \text{ and } sf_6(E_s) = 0.30$$

the final expression becomes

$$S_1 = 0.22 \ 10^{-8} \ \phi_f + 1.76 \ 10^{-8} \ \phi_s + D_\gamma \tag{10}$$

$$S_6 = 0.53 \ 10^{-8} \ \phi_f + 9.12 \ 10^{-8} \ \phi_s + D_\gamma \tag{11}$$

$$S_7 = 0.17 \ 10^{-8} \ \phi_f + D_\gamma \tag{12}$$

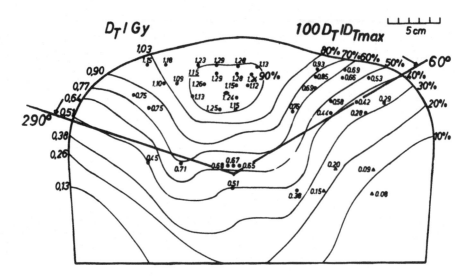

FIGURE 75.   Computer plan for fast neutron arc therapy using an A-150 body-shaped phantom. The underlined values refer to the total dose (D_T) isodose lines and the dots represent the measurement points. Measurement results are presented in Gy. (After Rassow, J., Advisory Group Meeting on Advances in Dosimetry for Fast Neutrons and Heavy Charged Particles for Therapy Applications, International Atomic Energy Agency, Vienna, June 14 to 18, 1982.)

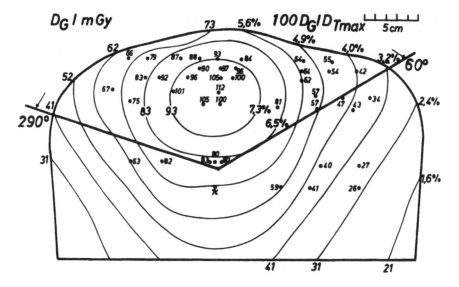

FIGURE 76.   The same treatment as in Figure 75. Only the gamma-ray dose component is presented. (After Rassow, J., Advisory Group Meeting on Advances in Dosimetry for Fast Neutrons and Heavy Charged Particles for Therapy Applications, International Atomic Energy Agency, Vienna, June 14 to 18, 1982.)

By using these equations, the fluence of fast and slow neutrons, as well as the gamma ray contribution, can be *approximately* determined in a clinical dosimetry situation. For a more comprehensive and critical analysis of the three-component model the reader is referred to Volume II, Chapter 2, Section IV.B.

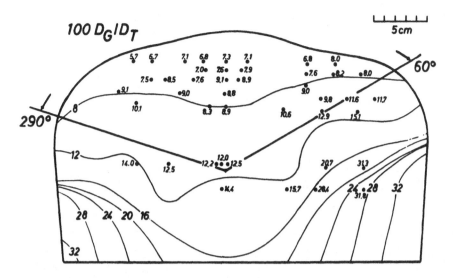

FIGURE 77.   The same treatment as in Figure 75. The gamma-ray dose fraction (D_G/D_T) is presented. (After Rassow, J., Advisory Group Meeting on Advances in Dosimetry for Fast Neutrons and Heavy Charged Particles for Therapy Applications, International Atomic Energy Agency, Vienna, June 14 to 18, 1982.)

Table 4
ISOTOPIC CONTENT OF
LIF TL MATERIALS

Phosphor	^{6}LiF (%)	^{7}LiF (%)
TLD-100	7.5	92.5
TLD-600	95.6	4.4
TLD-700	0.01	99.99

Table 5
AVERAGE
SENSITIVITY TO
^{60}CO GAMMA RAY

Material	Sensitivity cGy^{-1} mm^{-1}
TLD-100	11.4 ± 2.3
TLD-600	22.1 ± 3.5
TLD-700	20.6 ± 6.1

V. APPLICATIONS IN DIAGNOSTIC RADIOLOGY

TL dosimetry in diagnostic radiology has in the past been rather restricted to personnel dosimetry checks and local investigations such as in dental radiography, mammography, and chest radiography.[78-82] It has now achieved broader application, e.g., in beam quality studies and the performance of CT scanners.

A. Absorbed Dose Measurements in CT Scanner Beams

The Radiation Protection Institute (SSI) in Sweden has issued directions concerning the

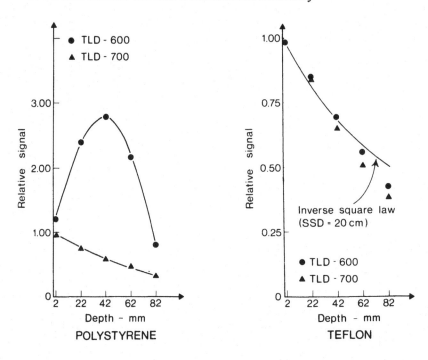

FIGURE 78. The relative, standardized depth response curves in polystyrene (left) and in teflon® with 25% carbon (right) for TLD-600 and TLD-700. (After Larsson, L., Alpsten, M., Lindskoug, B. A., and Sköldborn, H., *Nucl. Instrum. Methods*, 175, 189, 1980.)

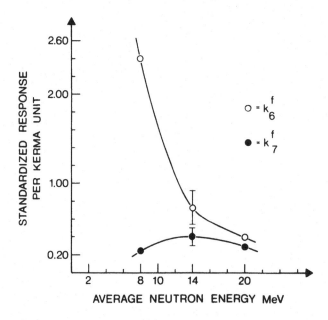

FIGURE 79. The fast neutron calibration factors k_6^f and k_7^f as a function of average neutron energy.

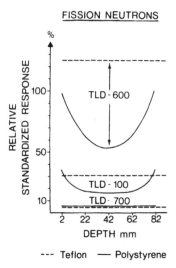

FIGURE 80. The relative standard-ized depth response curves in poly-styrene (   ) and teflon® with 25% C (− − −) for TLD-600, TLD-700, and TLD-100. Thermal neutron irradiation.

Table 6
CALIBRATION FACTORS FOR SLOW
(s) AND FAST (f) NEUTRONS

Material	Factor k_n^s	Factor k_n^f
TLD-100 (n = 1)	0.15 ± 0.01	0.48 ± 0.15
TLD-600 (n = 6)	0.16 ± 0.01	0.79 ± 0.24
TLD-700 (n = 7)	0.00	0.38 ± 0.12

use of CT scanners in radiological work. Among others, suggestions were made as to the physical dose measurements that should be carried out in order to check the quality and the safety of the equipment. Among the absorbed dose measurements the following were requested:

1. Dose profile and line integral for the clinical settings used at five specified positions in a cylindrical phantom
2. Dose profile at the center of the phantom for at least three of the slice thicknesses used
3. Dose profiles of sequential scans for different slice distances
4. Dose profile line integral at the center of the phantom as a function of the different parameters of X-ray exposure that affect the line integral
5. Line integral at the surface of the phantom as a function of the tube voltage setting used.

Lindskoug et al.[83] described a versatile method for carrying out these types of measurements using the continuous dosimeter readout by linear motion through a heated oven. The poly-

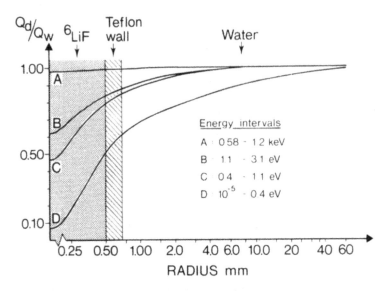

FIGURE 81.   The ratio (sf) as a function of the radius of a cylindrical water phantom for TLD-600 and various neutron energy intervals. Computed by ANISN program.[77]

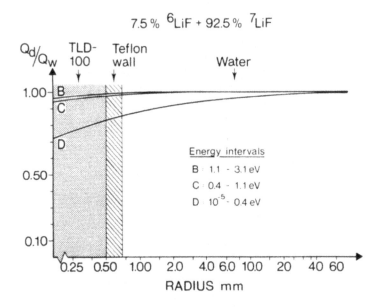

FIGURE 82.   The ratio (sf) for TLD-100 as a function of the radius of a cylindrical water phantom for various neutron energy intervals. Computed by ANISN program.[77]

styrene phantom recommended by SSI is shown in Figure 83. The thickness is 5 cm, the diameter is 20 cm, and the peripheral holes are at 10-mm depth below the surface. The continuous dosimeters are inserted half way into the holes. A CT radiograph of the phantom is shown in Figure 84. Note that the five dosimeters are clearly visible as small white dots.

The TLD reader (TLD-20) is on-line with a CBM computer. The data are collected by the computer and individual calibration factors are applied for each millimeter along each measuring probe.

The evaluation of the results involves the following steps:

Table 7
SELF-SHIELDING FACTORS
COMPUTED WITH ANISN

Energy interval	TLD-700	TLD-100	TLD-600
0.58—1.2 keV	1.0	1.00	0.98
1.1 —3.1 eV	1.0	0.98	0.72
0.4 —1.1 eV	1.0	0.96	0.63
10^{-5}—0.4 eV	1.0	0.78	0.30

FIGURE 83.   Polystyrene phantom recommended for CT-scanner beam-quality checks. The diameter is 20 cm and the thickness is 5 cm. Continuous dosimeters are inserted halfway into 2-mm holes drilled at 10-mm depth and at the center.

1.   Plotting of the primary data
2.   Integration of the primary peak
3.   Computation of the absorbed dose
4.   Plotting of the absorbed dose curve
5.   Listing of the peak-absorbed dose, the average dose, and the half-maximum width
6.   Computation of the line integral of the peak

As the readout of each dosimeter string is automatic and the computer is programmed to go through the data reduction and analysis automatically, many measurements can be made in a rather short time and with little personnel time.

1. Results

A typical plot of the primary data is shown in Figure 85. The corresponding absorbed

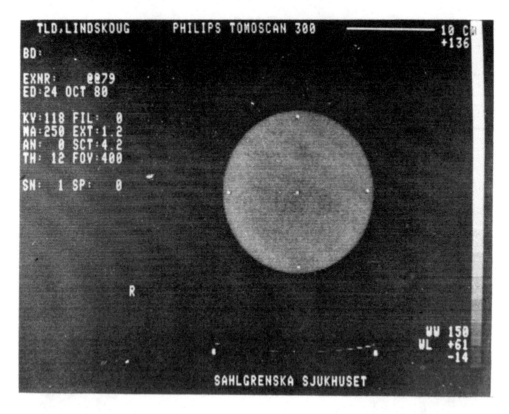

FIGURE 84.   CT radiograph of the phantom shown in Figure 83. Note that the dosimeters are clearly visible as small white dots.

dose data after application of the calibration factors are also presented in the figure. Values are given both for the periphery and the center of the cylindrical phantom block.

The axial resolution along the dosimeter is checked by exposure through a 1-mm slit collimator of lead, placed directly on top of the dosimeter, using a homogeneous 120 kVp X-ray field. The resulting readout plot shown in Figure 86 demonstrates that a 1-mm slit gives a 3-mm half-maximum width.

Figure 87 shows the measured peaks of a CT scan with slice thickness of 3 mm. The curves in Figure 88 show the line integral vs. the slice width for single and dual scans with the Tomoscan 300 (Philips®); 100% corresponds to 1100 mGy mm. It is obvious from this study that from the radiation protection point of view it is worthwhile to reduce the thickness of the slices without reducing the table increment.

B. Dosimetry at Interfaces

The absorbed dose in tissue in contact with materials of higher atomic number is raised over the electronic equilibrium-absorbed dose in the same photon fluence. This effect is due to the larger cross section for photoelectric absorption in the high Z material than in tissue. The effect is most pronounced in the low energy X-ray fields used for diagnostic radiology where high Z materials are used as contrast media. Barium (Z = 56) and iodine (Z = 53) are very often used in X-ray diagnostics. In dental radiology high Z materials like Hg (Z = 80) and Au (Z = 79) are often present in the beam. Alm Carlsson investigated the dose distribution in low Z materials in contact with materials of higher Z for 100 to 200 kVp X-ray beams using LiF-teflon® dosimeters.[85] The dosimeters are disks of 0.13×13 mm diameter. By using group calibration and avoiding high temperature annealing, a precision

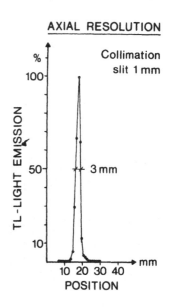

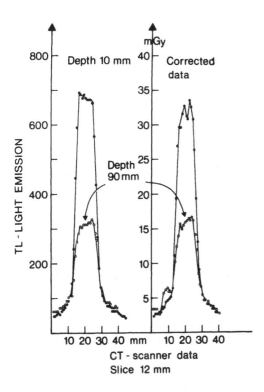

FIGURE 85. Primary measurement data in a CT scanner beam (left). A 1-mm integration interval is used. The corresponding absorbed dose data after application of individual (mm) calibration factors are shown to the right.

FIGURE 86. The response after exposure through a 1-mm slit collimator of lead, placed directly over the dosimeter string in a homogeneous 120 kV$_p$ X-ray field. The half-maximum value is 3 mm.

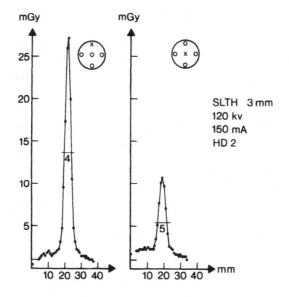

FIGURE 87. Typical results in a CT scanner beam with a slice width of 3 mm.

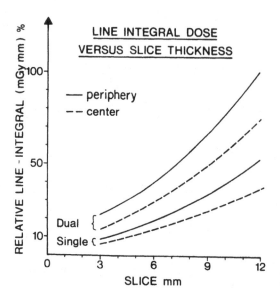

FIGURE 88. The line integral vs. the slice thickness for single and dual scans with Tomoscan 300. The data are normalized to slice width 12 mm and corrected for mAs and distance. The 100% value corresponds to 1100 mGy mm.

of 0.3% was achieved. The readout temperature was 310°C and cooling to room temperature was made reproducible within 1 min.

The depth doses in teflon®, with foils of Sn and Cu inserted 18 mm below the surface of the phantom, are illustrated in Figure 89. The solid curve is a reference depth dose curve for the homogeneous material. The dose distribution within the transition region is further analyzed in detail. Figure 90 shows the depth doses close to Pb-, Sn-, Cu-, and Al foils in mylar®.

Finally, Figure 91 summarizes the dose enhancement factor, defined as the ratio of the average dose close to a heterogeneity and the dose at the same point in electronic equilibrium. The enhancement ranges from a factor of 2 to a factor of 40. The curves are smoothed in order to eliminate the influence of the discontinuities from the K-edge emission.

VI. FUTURE CLINICAL APPLICATIONS OF THERMOLUMINESCENT DOSIMETRY

In spite of general concern about the harmful effects of ionizing radiation, the use of radiation sources seems to be constantly on the increase. New therapy sources are installed, nuclear medicine departments expand their use of radionuclides, and diagnostic departments continuously increase their investment in equipment associated with ionizing radiation. The need for clinical dosimetry, therefore, continues to grow.

The ever increasing demands for high precision radiotherapy has led to the construction of improved treatment units and improved therapy planning. These must all be checked by means of clinical dosimetry with precision and accuracy in accordance with the level of sophistication of the equipment and techniques. More sophisticated treatment techniques will certainly increase the need for more sophisticated dosimetric methods. It is already possible to envisage computerized dynamic therapy in which up to ten parameters will be varied during the therapy session, all guided by computer. This is called "conformation" therapy because it is possible to shape the isodose surfaces so that they conform to the

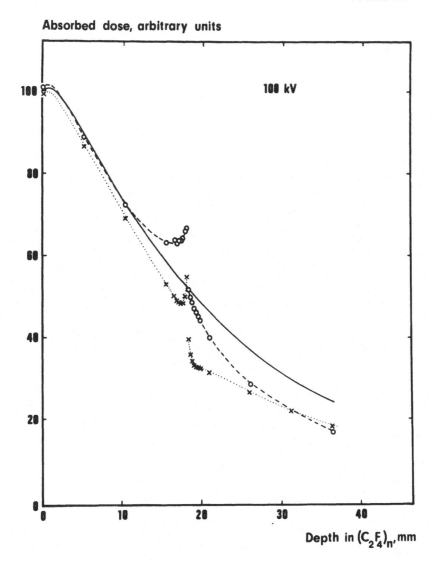

FIGURE 89. Depth dose in teflon® obtained with 100 kVp X-ray. Metal foils of SN (– – –) and Cu (⋯⋯⋯) are inserted 18 mm below the phantom surface. The solid curve is a reference depth dose curve for the homogeneous phantom. (After Alm Carlsson, G., *Acta Radiol.*, Suppl. 332, 1973.)

surface of the target volume. In fact, one is tempted to suggest that conformation therapy calls for "conformation" dosimetry in which the dosimeters would conform as much as possible to the curvature of the irradiated body or to the target volume. The continuous LiF-teflon® long-rod dosimeters seem very well suited for this purpose.

In diagnostic radiology the demands for high precision and reliable diagnosis are also leading to ever more sophisticated equipment and procedures. This is a field where TL dosimeters can contribute by monitoring the absorbed dose to the exposed populations and by beam quality determinations.

Target-seeking radioactive antibodies, used in radiotherapy, open a new field for in vivo measurements.[87] It is possible that the TL tube wall dosimeters can be of value in sampling and measuring the activities from inside the irradiated body.

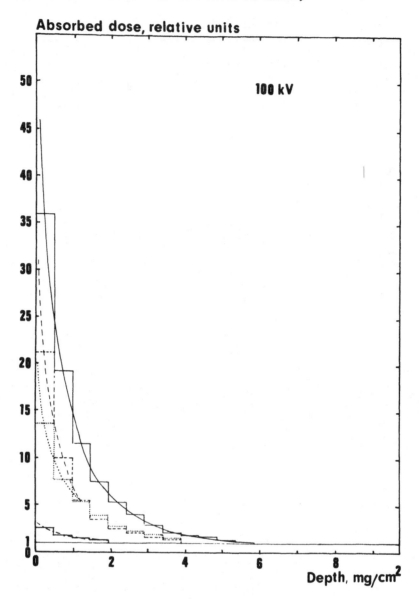

FIGURE 90.   Depth dose within the transition region in mylar, showing contribution from photoelectrons generated in an adjacent metal foil of Pb (———), Sn (– – –), Cu (·····), and Al (– . – . –). Energy 100 kV$_p$ X-ray. The continuous curves are fitted to the experimentally obtained histograms. (After Alm Carlsson, G., *Acta Radiol.*, Suppl. 332, 1973.)

In summary, the great simplicity of TL materials, especially their mechanical performance and geometric variability, makes them well suited for all aspects of clinical dosimetry work. More effort is required on the improvement of TL materials and their optimization in applications within the field of medical radiology.

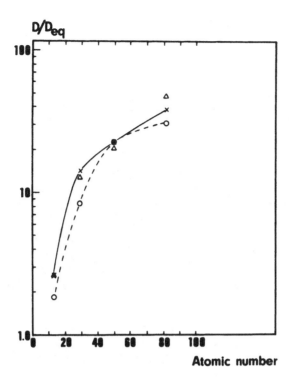

FIGURE 91. Dose enhancement factors in tissue, averaged over an 0.486 mg cm^{-2} layer close to metal foils of Pb, Sn, Cu, and Al, irradiated with primary photons generated at 100 kV ($\times$———$\times$) and 200 kV (o———o) potential. Experimental results of Kramer obtained with 60 kVp X-ray and with exoelectron-emitting dosimeters of BeO are included ($\triangle$).[84]

Appendix 1
CALIBRATION OF TL DOSIMETERS IN CLINICAL USE

The calibration must be made in accordance with the intended use of measurement. Let m be the measurement response, i.e., the TL light sum of the main peak, or peaks, of the TL glow curve after exposure to ionizing radiation; c the calibration response in the same beam; CD the calibration absorbed dose (Gy) in tissue at the position of the TL dosimeter when it is replaced by tissue; and f the calibration factor for the dosimeter in this beam. Then

$$f = CD/c \qquad (1)$$

and

$$MD = m\,f \qquad (2)$$

where MD is the measured absorbed dose in the beam.

This simple calculation has, of course, a very restricted validity. It can only be used if the dose (CD) is known beforehand at the point of measurement, the calibration and measurement are made in the same beam with the same geometry, and if the sensitivity of the measurement instrument is the same on both occasions. In reality it is necessary to proceed in a more generalized manner. The sensitivity of the TL material changes and the light sensitivity of the readout system changes. Corrections must be made for these discrepancies.

The light sensitivity of the readout system must be checked by a constant light source each day or in connection with each readout. Actually, one of the LiF dosimeters can be used as a constant light source if it is always irradiated to the same low dose, used every day and stored in a refrigerator when not in use.

The relative light-correction factor (f_L) is defined as the ratio of the light sensitivity of the system during calibration and the corresponding sensitivity at the measurement. Then

$$MD = m \, f \, f_L \tag{3}$$

Routinely used TL materials will change their sensitivity depending on the accumulated dose, the time of storage, and the temperature history. Thus the calibration factor (f) must be determined repeatedly. Instead of repeating the calibration for each one of the radiation sources, it is convenient to choose one of the sources as a reference source. ^{60}Co γ-rays are often used as a reference source. Let r be the reference response; f_r the reference source calibration factor; RD the reference absorbed dose (Gy); and $\eta_{c,r}$ the relative TL-response. Then

$$f_r = RD/r \tag{4}$$

and

$$\eta_{c,r} = f/f_r = \frac{CD \, r}{c \, RD} \tag{5}$$

The reference source relative TL response ($\eta_{c,r}$) must be established for all radiation sources in use. The measured dose is then computed according to

$$MD = m \, f_r \, \eta_{c,r} \, f_L \tag{6}$$

It would be ideal if one initial calibration to determine $\eta_{c,r}$ for all dosimeters and radiation sources was sufficient. Although $\eta_{c,r}$ is quite constant for high energy electrons and photons, the sensitivity of the dosimeters in the various reference beams may change when the accumulated dose increases, particularly if the LET of the various beams differs. The reference irradiator can thus be used to monitor sensitivity only up to a certain accumulated dose level, the value of which must be determined for each dosimeter batch and source.

If the sensitivity is found to change rapidly, the reference irradiation must be frequently repeated. The extreme is to repeat the reference irradiation after each measurement and apply the corresponding correction each time the dose is computed.[37] Note that the sensitivity increase with accumulated dose is less for high LET radiation than for low LET radiation. A recommended starting value of accumulated dose limit, for repeated reference irradiations, is 10 Gy and a complete new calibration should be carried out after 100 Gy.

Appendix 2
GROUPING OF TL-DOSIMETERS

Calibration of a single dosimeter was considered in Appendix 1. Usually, many dosimeters are in circulation at the same time and a system for group calibration may be practical.

In a group of (n) matched dosimeters the individual response for dosimeter i (m_i) and the average response of the group ($\bar{m}$) give an approximatively constant ratio.

$$g_i = \overline{m}/m_i, \text{ where } \overline{m} = \frac{1}{n} \sum_{i=1}^{n} m_i \tag{1}$$

as long as the dosimeters are used in parallel and given about the same accumulated dose and thermal history.

Considerable calibration work can be saved if a few of the dosimeters are taken as a reference group to be calibrated in connection with each measurement. Let $\overline{r}$ be the average reference source response of the reference group; $\overline{c}$ the average calibration response of the reference group; m_i the individual dosimeter (i) measurement response; and g_i the single-dosimeter group correction factor defined above. Then

$$MD = m_i \, g_i \, f_r \, \eta_{c,r} \, f_L \tag{2}$$

where

$$f_r = RD/\overline{r} \text{ and } \eta_{c,r} = \frac{CD}{\overline{c}} \frac{\overline{r}}{RD} \tag{3}$$

f_r is the current reference source calibration factor for the reference group and $\eta_{c,r}$ the relative TL response, calibration source to reference source, for the entire group (n). This formula gives the average dose to the group had all the dosimeters been irradiated in the same beam.

It is also well known that in a group of dosimeters that are repeatedly irradiated and read out, the individual dosimeter will vary more than the group average response. Thus the method of group calibration gives a random variation smoothing.

Appendix 3
CALIBRATION OF CONTINUOUS DOSIMETERS

The most direct way of using and calibrating continuous dosimeters is to treat each integration interval length, along the dosimeter string, as an independent individual dosimeter and calibrate each dosimeter for all possible integration interval lengths. This means that computer mass-storage is needed to keep track of all the dosimeters and the various calibration factors for each integration interval. A new data file is created for each dosimeter and integration interval. Let L be the length of a dosimeter (mm); s the integration interval (mm); m_i the measurement response for integration interval i, $1 \leq i \leq N$, where $N = L/s$, the number of integration intervals along the dosimeter, c_i the calibration response for integration interval number i; and f_i the calibration factor for integration interval i, MD_i the calculated absorbed dose for interval i. Then

$$f_i = CD/c_i \tag{1}$$

and the dose for interval number i is computed according to

$$MD_i = m_i \, f_i \, f_L, \text{ where } i = 1,2,3,\ldots,N. \tag{2}$$

In the same manner as was described in Appendix 1, a reference irradiator will simplify the calibration work. Let r_i be the reference response for integration interval number i; RD the reference absorbed dose (Gy) for interval i; $f_{r,i}$ the reference calibration factor for interval i; and $\eta_{c,r}$ the relative TL response calibration to reference irradiation. Then

$$\eta_{c,r} = \frac{CD \, r_i}{c_i \, RD} \tag{3}$$

$$f_{r,i} = \frac{RD}{r_i} \qquad (4)$$

and

$$\eta_{c,r} = f_i/f_{r,i} \qquad (5)$$

The measured dose for interval number i (MD_i) is then given according to

$$MD_i = m_i \, f_{r,i} \, \eta_{c,r} \, f_L \qquad (6)$$

It is presumed that $\eta_{c,r}$ is constant for all intervals along a dosimeter, whereas $f_{r,i}$ and f_i may change from one end of a dosimeter to the other. Thus it is rarely possible to utilize the average value of all the integration interval data along a dosimeter even if the irradiation is known to be homogeneous. It may, however, be quite useful to apply a smoothing procedure in order to reduce the random variation and also the number of data, particularly if the integration interval is small, because the standard deviation of the measurement increases when the integration interval becomes shorter.

SMOOTHING

A simple method of smoothing that does not violate the primary data can be used to reduce the number of data and the random variation. Let 2 K be the length over which smoothing averaging is started (mm), s the integration interval (mm); 2 k the number of integration intervals (= number of data) over the length 2 K; and $\bar{r}_i$ the reference response for integration interval number i after smoothing. Then

$$k = K/s \qquad (7)$$

For i = k, 2k, 3k, ... , L/ks, let

$$\bar{r}_i = \frac{1}{2k+1} \sum_{n=i-k}^{i+k} r_n, \text{ limited by } \begin{cases} i + k < L/s \\ i - k > 0 \end{cases} \qquad (8)$$

which is the same limitation as $0 < n < L/s$.

If any of the end values of r_n differ from the average over the interval (2 k) by more than a preset fraction δ, then k is reduced by 1 and the procedure repeated. Thus, if

$$\frac{|\bar{r}_i - r_{i-k}|}{\bar{r}_i} > \delta \text{ or } \frac{|\bar{r}_i - r_{i+k}|}{\bar{r}_i} > \delta \qquad (9)$$

then k is given the value k - 1 and the averaging is repeated until the fraction δ is superior. Then the current value of $\bar{r}_i$ is taken as the response for the integration interval number i. The next smoothed response value will correspond to integration interval number i + k and so on.

For the calculation of absorbed dose, let $\bar{c}_i$ be the smoothed calibration response over the integration interval number i; $\bar{m}_i$ the smoothed measurement response over interval i; and f_i the calibration factor for interval i. Then

$$f_i = CD/\bar{c}_i \qquad (10)$$

and

$$MD_i = \overline{m}_i \, f_i \, f_L \tag{11}$$

where

$$i = k, 2k, 3k, \ldots, L/ks$$

If a reference irradiator is used and $\overline{r}_i$ is the smoothed reference response for interval i; $\eta_{c,r}$ the relative TL response; $f_{r,i}$ the reference calibration factor for interval i, then

$$f_{r,i} = RD/\overline{r}_i \tag{12}$$

$$\eta_{c,r} = \frac{CD \, \overline{r}_i}{\overline{c}_i \, RD} \tag{13}$$

and

$$MD_i = \overline{m}_i \, f_{r,i} \, \eta_{c,r} \, f_L, \; i = k, 2k, 3k, \ldots, \leqslant L/ks \tag{14}$$

Note that the fraction δ must be determined for each measurement situation in order not to violate the primary data.

Appendix 4
SURFACE MAXIMUM RATIO

The surface maximum ratio (SMR) relates a TL measurement on the surface of a tissue-equivalent phantom to a simultaneous TL measurement at the depth of maximum build-up.[46] Let PDD(d,A) be the percentage depth dose at depth d and field area A and m(d,A) the TL response at depth d, area A. Then, by definition

$$SMR(A) = m(0,A)/m(t,A) \tag{1}$$

where t is the depth of the dose maximum. The absorbed dose at depth d is then given by the expression

$$D(d,A) = m(0,A) \frac{PDD(d,A)}{SMR(A) \, 100} \, f_r \, \eta_{c,r} \, f_L \tag{2}$$

where m(O,A) is the TL response at the surface of the patient treated with a field of area A.

SMR is a nonuniversal empirical function that depends on the TL-material used and the geometry of application, etc. SMR must therefore be locally determined and tabulated for all sources and field sizes. Used with care it is of great help in simplifying the calculations.

Appendix 5
ALGORITHM FOR CALCULATION OF FLATTENING FILTER

Let SSD be the source-skin distance; SFD the source-filter distance; D the relative sum absorbed dose from the anterior and the posterior field, D_o the flat level to which the dose distribution is shaped by the filter, N the number of the filter element, $N = 1,2,3,...,42$ $S(N)$ the thickness of filter element N. $X(N)$ the distance from the central ray in the midplane of the body, B_o the body thickness along the central ray, average of the AP and the PA positions, Y the curve-tracer coordinate corresponding to the graph level above tracer-table-origin (O); and μ_{eff} the broad beam attenuation coefficient for the aluminum slabs used in the filter.

Each of the 42 elements making up the filter is 5-mm wide. The filter base plate is mounted at 665 mm from the focus of the therapy machine. Then

$$Z = SSD + B_o/2 \tag{1}$$

$$N = (105 - SFD\ X(N)/Z)/5 \tag{2}$$

$$D = c_1 Y + c_2 \tag{3}$$

$$S(N) = \ln(D/D_0)/\mu_{eff} \tag{4}$$

where c_1 and c_2 are constants for the tracer coordinate system.

The filter correction factor for the monitor setting is calculated as

$$FAC = 100/D_0 \tag{5}$$

CALCULATION OF MIDPLANE DOSE

If the patient refuses to take the esophagus measurement, transmission measurements must be made on the entrance and exit sides of the body. Let D_{mp} be the calculated midplane absorbed dose; D_i be the measured dose on the entrance side with full build-up; D_e the measured dose on the exit side; d_{max} the depth of absorbed dose maximum. Then

$$D_e = D_i\ (SSD/(SSD + d_m))^2\ \exp(-a) \tag{6}$$

$$a = \ln(D_i(SSD/(SSD + d_m))^2/D_e) \tag{7}$$

and

$$D_{mp} = D_e\ \exp(0.5\ a) \tag{8}$$

ACKNOWLEDGMENTS

The authors are grateful to all the colleagues who contributed with their published materials in the form of figures as well as text. We also want to acknowledge the assistance of Dr. Gunilla Kjellberg for support and discussion of the clinical work. We wish to express our appreciation for the cooperation with the staff at the Computer Tomograph at Rtg I, Sahlgren Hospital.

Our thanks are due to Maria Elsby for typing the manuscript, Viveka Norlin for the drawings, and Ole Roos for the photographs.

REFERENCES

1. **Spiers, F. W. and Zanelli, G. D.**, Experimental methods of dosimetry in bone with special reference to dose in trabecular bone, in Proc. 11th Int. Congr. Radiol., Rome, 1965, 1809.
2. **Almond, P. R., Wright, A., and Lontz, J. F., II**, Use of lithium fluoride thermoluminescent dosimeters with high-energy electron beams, in Proc. Symp. Solid State and Chemical Radiation Dosimetry in Medicine and Biology. STI/PUB/138, IAEA, Vienna, 1967, 53.
3. **Gooden, D. S. and Brickner, T. J.**, Thermoluminescence dosimetry for clinical use in radiation therapy, in Proc. 3rd Int. Conf. Luminescence Dosimetry, AEC, Risö, Denmark, 1971, 793.
4. **Rudén, B. I.**, Two years experience of clinical thermoluminescence dosimetry at the radiumhemmet, Stockholm, in Proc. 3rd Int. Conf. Luminescence Dosimetry, Rep. No. 249, AEC, Risö, Denmark, 1971, part 2, 781.
5. **Joelsson, I., Rudén, B. I., Costa, A., Dutreix, A., and Rosenwald, J. C.**, Determination of dose distribution in the pelvis by measurement and by computer in gynecologic radiation therapy, *Acta Radiol. Ther. Phys. Biol.*, 11, 289, 1972.
6. **Suntharalingam, N. and Mansfield, C. M.**, The use of thermoluminescence dosimeters in high energy electron and photon beam clinical dose measurements, in Proc. 4th Int. Conf. Luminescence Dosimetry, Institute for Nuclear Physics, Krakow, 1974, 859.
7. **Mansfield, C. M. and Suntharalingam, N.**, The usefulness of thermoluminescence dosimetry in clinical radiation therapy, in Proc. Symp. Biomedical Dosimetry, IAEA, Vienna, 1975, 335.
8. **Wall, B. F., Bird, P. D., Fisher, E. S., Hudson, A., and Paynter, G.**, Doses to patients from pantomographic and conventional dental radiography, *Br. J. Radiol.*, 52, 727, 1979.
9. **Wall, B. F., Green, D. A. C., and Veerappan, R.**, The radiation dose to patients from EMI brain and body scanners, *Br. J. Radio.*, 52, 189, 1979.
10. **Noel, A., Aletti, P., Bey, P., Hoeffel, J. C., Hoffstetter, S., and Schoumacher, P.**, Thermoluminescent dosimetry, our experience with clinical applications, in Proc. 6th Int. Conf. Solid State Dosimetry, *Nucl. Instrum. Methods*, 175, 208, 1980.
11. **Sievert, R. M.**, Eine Methode zur Messungen von Röntgen-, Radium und Ultrastrahlung nebst einige Untersuchungen über die Anvendbarkeit derselben in der Physik und der Medizin, *Acta Radiol.*, Suppl. 14, 1932.
12. **Benner, S.**, An instrument for calculating roentgen ray doses from condenser chamber readings. *Acta Radiol.*, 27, 243, 1946.
13. **Sköldborn, H.**, On the design, physical properties and practical application of small condenser ion chambers. *Acta Radiol.*, Suppl. 187, 1959.
14. **Lidén, K.**, Depth dose measurements in the esophagus in roentgen rotation therapy, *Acta Radiol.*, 30, 64, 1948.
15. **Dahl, O. and Vikterlöf, K. J.**, Dose distributions in arc therapy in the 200 to 250 kV range, *Acta Radiol.*, Suppl. 171, 1958.
16. **Benner, S., Ragnhult, I., and Gerbert, G.**, Miniature ionization chambers for measurements in body cavities, *Phys. Med. Biol.*, 4, 26, 1959.
17. **Hultberg, S., Dahl, O., Thoreaus, R., Vikterlöf, K. J., and Walstam, R.**, Kilocurie cobalt 60 therapy at the radiumhemmet, *Acta Radiol.*, Suppl. 179, 1959.
18. **Nordberg, U. B.**, Determination of the tumour dose by transmission measurements in roentgen rotation treatment of the esophagus, *Acta Radiol. Ther.*, 7, 401, 1968.
19. **Joelsson, I. and Bäckström, A.**, Dose rate measurements in bladder and rectum, *Acta Radiol. Ther.*, 8, 343, 1969.
20. **Svahn-Tapper, G. and Landberg, T.**, Mantle treatment of Hodgkins disease with cobalt 60, *Acta Radiol. Ther.*, 10, 1, 1971.
21. **Daniels, F., Boyd, C. A., and Saunders, D. F.**, Thermoluminescence as a research tool, *Science*, 117, 343, 1953.
22. **Bjärngard, B.**, Use of manganese and samarium activated calcium sulphate in thermoluminescence dosimetry, in Proc. 1st Int. Conf. Luminescence Dosimetry, U.S. A.E.C. CONF 650637, NTIS, Springfield, Va., 1967, 195.
23. **Bjärngard, B. E. and Jones, D.**, Thermoluminescent dosimeters of LiF and CaF_2:Mn incorporated in teflon, in Proc. Symp. Solid State and Chemical Radiation Dosimetry in Medicine and Biology, IAEA, Vienna, 1967, 99.
24. **Benner, S., Johansson, M., and Lindskoug, B.**, Radiation dosimetry in small cavities using thermoluminescence in lithium fluoride powder, *Nature (London)*, 212, 5061, 1966.
25. **Lindskoug, B., Johansson, J. M., Karlsson, R., and Kellgren, R.**, Measuring device for thermoluminescent dosimetry, *J. Sci. Instrum.*, 44, 939, 1967.
26. **Carlsson, C. A.**, Thermoluminescence of LiF. Dependence of thermal history, *Phys. Med. Biol.*, 14, 107, 1969.

27. **Mårtensson, B. K. A.**, Thermoluminescence of LiF. A statistical analysis of the influence of pre-annealing on the precision of measurement, *Phys. Med. Biol.*, 14, 119, 1969.

28. **Rudén, B. I.**, Some Applications of Thermoluminescence Dosimetry in Medical and Health Physics, Ph.D. thesis, University of Stockholm, Stockholm, Sweden, 1975.

29. **Lindskoug, B. and Bengtsson, B. E.**, Automated thermoluminescence reader. I. Technical construction and function, *Acta Radiol. Ther. Phys. Biol.*, 14, 195, 1975.

30. **Cameron, J. F., Suntharalingam, N., and Kenney, G. N.**, *Thermoluminescent Dosimetry*, The University of Wisconsin Press, Madison, 1968.

31. **Becker, K.**, *Solid State Dosimetry*, CRC Press, Boca Raton, Fla., 1973.

32. **McKinley, A. F.**, *Thermoluminescence Dosimetry*, Medical Physics Handbook 5, Adam Hilger Ltd., Bristol, 1981.

33. **Oberhofer, M. and Scharmann, A., Eds.**, *Applied Thermoluminescence Dosimetry*, Adam Hilger Ltd., Bristol, 1981.

34. **Horowitz, Y. S.**, The theoretical and microdosimetric basis of thermoluminescence and applications to dosimetry, *Phys. Med. Biol.* 26, 765, 1981.

35. **Zimmerman, D. W., Rhyner, C. R., and Cameron, J. R.**, Thermal annealing effects on the thermoluminescence of LiF, *Health Phys.*, 12, 525, 1966.

36. **Benner, S., Johansson, J. M., Lindskoug, B., and Nyman, P. T.**, A miniature LiF dosimeter for in vivo measurements, Proc. Symp. Solid State and Chemical Radiation Dosimetry in Medicine and Biology, IAEA, Vienna, 1967, 65.

37. **Lindskoug, B.**, Automated thermoluminescence reader. II. Experiments and theory, *Acta Radiol. Ther. Phys. Biol.*, 14, 347, 1975.

38. **Lindskoug, B. A.**, Continuous cylindrical dosimeters in TSL dosimetry, Digest of 12th Int. Conf. Med. Biol. Eng./5th Int. Conf. Med. Phys., Jerusalem, 1975, part 4, 63.1.

39. **Horowitz, Y. S., Fraier, I., Kalef-Ezra, J., Pinto, H., and Goldbart, Z.**, Non-universality of the TL-LET response in thermoluminescent LiF. The effect of batch composition, *Phys. Med. Biol.*, 24, 1268, 1979.

40. **Suntharalingam, N. and Cameron, J. R.**, Thermoluminescent response of lithium fluoride to radiations with different LET, *Phys. Med. Biol.*, 14, 397, 1969.

41. **Chan, F. K. and Burlin, T. E.**, The energy-size dependence of the response of thermoluminescent dosimeters to phantom irradiation, *Health Phys.*, 18, 325, 1970.

42. **Paliwal, B. R. and Almond, P. R.**, Applications of cavity theories for electrons to LiF dosemeters, *Phys. Med. Biol.*, 20, 547, 1975.

43. **O'Brien, K.**, Monte Carlo calculations of the energy response of lithium dosimeters to high energy electrons (>30MeV), *Phys. Med. Biol.*, 22, 836, 1977.

44. **Carlsson, C. A., Mårtensson, B. K. A., and Alm Carlsson, G.**, High precision dosimetry using thermoluminescent LiF, Proc. 2nd Int. Conf. Luminescence Dosimetry, U.S. A.E.C. CONF-680920, NTIS, Springfield, Va., 1968.

45. **Lindskoug, B. A.**, Further considerations on the use of continuous cylindrical dosimeters in TSL dosimetry, *Nucl. Instrum. Methods*, 175, 89, 1980.

46. **Bagne, F.**, Clinical use of thermoluminescent dosimeters in supervoltage, X-ray therapy, *Radiology*, 119, 480, 1976.

47. **Busch, M.**, Optimized intracavitary and interstitial treatment with a high dose rate afterloading system. First clinical results, presented at the 21st Annu. Meet. Am. Soc. Ther. Radiol., New Orleans, 1979.

48. **Lindskoug, B. A.**, Treatment planning for computerized interstitial afterloading, 1st Annu. Meet. Eur. Soc. Ther. Radiol. Oncol., London, 1982.

49. **Johansson, J. M., Lindskoug, B., and Nyström, C.**, Pelvic dosimetry during radiotherapy of carcinoma of the cervix uteri, *Acta Radiol. Ther. Phys. Biol.*, 8, 360, 1969.

50. **Lindskoug, B. A.**, Accepted treatment plan versus practical treatment, Proc. 6th Int. Conf. Use of Computers in Radiation Therapy, Rosenow, U., Ed., Göttingen, Germany, 437, 1977.

51. **Lindskoug, B. A. and Notter, G.**, Design of flattening filters based on intracavitary absorbed dose measurements in external radiation therapy, *Br. J. Radiol.*, 53, 976, 1980.

52. **Ragnhult, I., Lindskoug, B., and Hultborn, A.**, Dosimetric investigation of postoperative irradiation of regional lymph nodes in mammary carcinoma, *Acta Radiol.*, 313 (Suppl.), 135, 1972.

53. **Whitton, J. T.**, New values for epidermal thickness and their importance, *Health Phys.*, 24, 1, 1973.

54. **Di Paola, M., Bianchi, M., and Baarli, J.**, Lens opacification in mice exposed to 14-MeV neutrons, *Radiat. Res.*, 73, 340, 1978.

55. **Charles, M. W. and Brown, N.**, Dimensions of the human eye relevant to radiation protection, *Phys. Med. Biol.*, 20, 202, 1975.

56. **Jetne, V.**, Eye lens dose measurements, Digest of the 9th Nordic Meet. Clin. Phys., S-413 45, Roos, B., Ed., Sahlgren Hospital, Gothenburg, Sweden, 1977.

57. **Lindskoug, B. and Hultborn, A.**, Tissue heterogeneity in the anterior chest wall and its influence on radiation therapy of the internal mammary lymph nodes, *Acta Radiol. Ther. Phys. Biol.*, 15, 97, 1976.

58. **Lam, W.-C., Lindskoug, B. A., Order, S. E., and Grant, D. G.**, The dosimetry of ^{60}Co total body irradiation, *Int. J. Radiat. Oncol. Biol. Phys.*, 5, 905, 1979.

59. **Lam, W.-C., Order, S. E., and Thomas, E. D.**, Uniformity and standardization of single and opposing cobalt 60 sources for total body irradiation, *Int. J. Radiat. Oncol. Biol. Phys.*, 6, 245, 1980.

60. **Aget, H., VanDyk, J., and Leung, M. K.**, Utilization of a high energy photon beam for whole body irradiation, *Radiology*, 123, 747, 1977.

61. **Ssengabi, J.**, Studies of Photon Beam Properties in Therapeutic Applications, Ph.D. thesis, University of Stockholm, Stockholm, Sweden, 1978.

62. **Nilsson, B. and Schnell, P.-O.**, Build-up effects at air cavities measured within thin thermoluminescent dosimeters, *Acta Radiol. Ther. Phys. Biol.*, 15, 427, 1976.

63. **Bertilsson, G.**, Electron Scattering Effects on Absorbed Dose Measurements with LiF-Dosimeters, Ph.D. thesis, University of Lund, Lund, Sweden, 1975.

64. **Kastner, J., Hukkoo, R., Oltman, B. G., and Dayal, Y.**, Thermoluminescent internal beta-ray dosimetry, *Radiat. Res.*, 32, 625, 1967.

65. **Greitz, U. and Rudén, B. I.**, Calibration of LiF teflon rods for internal beta-ray dosimetry, *Phys. Med. Biol.*, 17, 193, 1972.

66. **Schulz, U.**, Die Optimierte Interstitielle afterloading-therapy, in Kombinierte Chirurgische und Radiologische Therapie maligner Tumoren, Urban – Schwarzenberg, Munich, 1981.

67. **Quast, U. and Borman, U.**, Experimentelle und Klinische Leukämie und Tumorforschung an der Universität Essen-Gesamthochschule, Medizinische Physik 80, Heidelberg, 2, 495, 1980.

68. **Lindskoug, B. A.**, Restricted optimization in afterloaded interstitial brachytherapy, 23rd Annu. Meet. American Association of Physicists in Medicine, Boston, 1981.

69. **Lindskoug, B. A.**, Restriction Optimization in Computerized Interstitial Brachytherapy, World Congress on Medical Physics and Biomedical Engineering, Hamburg, 1982.

70. **Lindskoug, B. A.**, TL-dosimeters by LiF-powder imbedded in the wall of teflon tubes, 4th Int. Conf. Med. Phys., Ottawa, Canada, 1976.

71. **Rassow, J.**, TLD-300 detectors for separate measurements of total and gamma absorbed dose distributions of single, multiple, and moving field neutron treatments — a new method of clinical dosimetry for fast neutron therapy, Advisory Group Meeting on Advances in Dosimetry for Fast Neutrons and Heavy Charged Particles for Therapy Applications, IAEA, Vienna, June 14 to 18, 1982.

72. **Lucas, A. C. and Kapsar, B. M.**, The thermoluminescence of thulium doped calcium fluoride, 5th Int. Conf. Luminescence Dosimetry, Sao Paolo, Brazil, 1977.

73. **Temme, A., Rassow, J., and Meissner, P.**, A new thermoluminescent dosimetry procedure using TLD-300 detectors for clinical dosimetry in mixed neutron-gamma ray field, 4th Symp. Neutron Dosimetry, München, 1981, in Radiation Protection, Commission of the European Communities, EUR 7448, 1981, 2, 433.

74. **Maier, E., Hüdepohl, G., Meissner, P., and Rassow, J.**, Erste Dosimetrie- und Strahlenschutzmessergebnisse an dem Isozentrischen Zyklotron-Neutronentherapiegerät CIRCE, Medizinische Physik '78, Hüthig-Verlag, 1978, 293.

75. **Lindskoug, B. A.**, Liaison measurements of fast neutrons by TL-dosimeters, Proc. Int. Conf. Appl. Phys. Med. Biol., Trieste, Italy, IAEA, Vienna, 1982.

76. **Horowitz, Y. S., Dubi, A., and Ben Shahar, B.**, Self-shielding factors for TLD-600 and TLD-100 in an isotropic flux of thermal neutrons, *Phys. Med. Biol.*, 21, 976, 1976.

77. **Engle, W. W., Jr.**, User Manual for ANISN, Rep. No. K-1693, Oak Ridge Gaseous Diffusion Plant, Oak Ridge, Tenn., 1967.

78. **Vacirca, S. J. and Thompson, D. L.**, Dose outside useful beam in water and realistic phantoms exposed to chest radiographic procedures, *Health Phys.*, 23, 533, 1972.

79. **Vacirca, S. J., Thompson, D. L., Pasternack, B. S., and Blatz, H.**, A film-thermoluminescent dosimetry method for predicting body doses due to diagnostic radiography, *Phys. Med. Biol.*, 17, 71, 1972.

80. **Wall, B. F., Fisher, E. S., Paynter, R., Hudson, A., and Bird, P. D.**, Dose to patients from pantomographic and conventional dental radiography, *Br. J. Radiol.*, 52, 727, 1979.

81. **Watson, P.**, A survey of radiation doses to patients in mammography, *Br. J. Radiol.*, 50, 745, 1977.

82. **Langmead, W. A., Wall, B. F., and Palmer, K. E.**, An assessment of lithium borate thermoluminescent dosimeters for the measurement of doses to patients in diagnostic radiology, *Br. J. Radiol.*, 49, 956, 1976.

83. **Lindskoug, B. A., Arvidsson, B., and Quiding, L.**, Dose profile measurements in a CT-scanner beam in Proc. 17th Int. Congr. Radiol., Brussels, 1981.

84. **Kramer, J.**, Exoelektronen-Dosimetre für Roentgen- und Gammastrahlen, *Z. Angew. Phys.*, 20, 411, 1966.

85. **Alm Carlsson, G.**, Dosimetry at interfaces. Theoretical analysis and measurements by means of thermoluminescent LiF, *Acta Radiol.*, Suppl. 332, 1973.

86. **Larsson, L., Alpsten, M., Lindskoug, B. A., and Skoldborn, H.,** The response of LiF dosimeters in the radiation field of a 14 MeV neutron generator, *Nucl. Instrum. Methods,* 175, 189, 1980.
87. **Order, S. E.,** Monoclonal antibodies: potential role in radiation therapy and oncology, *Int. J. Radiat. Oncol. Biol. Phys.,* 8, 1193, 1982.

Chapter 3

THERMOLUMINESCENCE APPLIED TO AGE DETERMINATION IN ARCHAEOLOGY AND GEOLOGY

V. Mejdahl and A. G. Wintle

TABLE OF CONTENTS

I. INTRODUCTION

The TL emission characteristic of numerous minerals has been known and studied for more than 150 years.[1] However, the cause of the emission remained obscure until 1905 when it was realized that TL in minerals could be induced by radiation from naturally occurring radionuclides.[2] In later studies[3,4] the correlation of the natural TL signal with radioactive impurities was demonstrated for a large number of common rock-forming minerals.

The possibility of using TL for dating heated materials was proposed by Daniels et al.[5] The first announcement of results obtained for archaeological samples was made by Kennedy and Knopf at a meeting of the American Association for the Advancement of Science in 1960, the same year that W.F. Libby was awarded the Nobel Prize for the development of the radiocarbon dating technique. They also reported on the application of the TL technique to young lava flows. The same year Grögler et al.[6] published the first paper that dealt with the application of TL to the dating of pottery. As a result of these preliminary studies, several more laboratories around the world became interested in the technique. In the early 1960s there were a few preliminary publications on the TL dating of pottery emphasizing the potential of the technique, but also reporting that it was more complex than had initially been thought.[7-9]

Most of the TL studies up until 1966 were concerned with applications to geological dating and stratigraphy and this work was reported at a conference held in Spoleto, Italy in 1966. The proceedings were published in 1968 in a book entitled *Thermoluminescence of Geological Materials*,[10] which also contained an extensive bibliography assembled by Grögler covering the application of TL to archaeology, geology, mineralogy, and meteorites. Most of the geological studies were connected with dating pre-Quaternary formations (with little success) or with trying to develop TL as a technique for use in uranium exploration. The following 10 years saw a greater emphasis on the application of TL to archaeological dating[11-13] and involved detailed dosimetry studies, first on ceramics and then on other materials such as burnt stones. Paralleling this, studies have been carried out on more recent geological materials including Quaternary volcanic products, calcitic deposits such as stalagmites, and most recently marine and terrestrial sediments.

II. THERMOLUMINESCENCE IN NATURE

A. Thermoluminescent Minerals

Most rocks contain at least one mineral that will give a TL signal when it is heated after irradiation. Since the early 1930s many studies have reported on the TL properties of such common minerals as calcite, dolomite, feldspars, quartz, apatite, and zircon. McDougall[14] summarized some of these reports. Clay minerals, such as illites and kaolinites, also exhibit TL.[15]

Nishita et al.[16] reported an attempt to classify minerals on the basis of their natural TL, but found a wide range of TL response for the 15 soil and 73 mineral and rock samples they studied. Other silicate structures, such as flint, chert, and obsidian, have received attention more recently since they are often the only datable material on early archaeological sites. The current status of TL studies on minerals and rocks has been surveyed by Sankaran et al.[17]

These different minerals often give rise to characteristic glow curves, although the peak structure and/or emission spectra may vary depending on the impurity concentration in the crystal. Indeed these variations have been used to characterize obsidian sources and thus provide a way of tracing native Indian trade routes in North America.[18] Similar studies on quartz have been used to trace sources of sedimentary rock in Belgium.[19] In most studies the glow curves are obtained by heating at a constant rate from room temperature to about

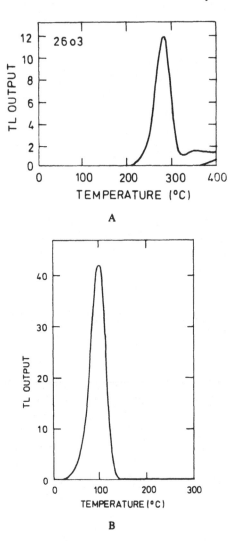

FIGURE 1.   (A) Typical natural glow curve for calcite. Heating rate is 5°C sec^{-1}. (After Wintle, A. G., *Can. J. Earth Sci.*, 15, 1977, 1978.) (B) Typical glow curve showing the 110°C peak of quartz. Heating rate is 8°C sec^{-1}.

500°C. Above this temperature the incandescence from the sample itself, as well as the heating plate, increases rapidly and is predominant in the photomultiplier output. Even to reach 500°C it is necessary to use an infrared absorbing filter and a blue color filter (Corning 7-59 or 5-58), as well as a photomultiplier (PM) tube with a low red response, such as an EMI 9635. Typical heating rates for dating applications range from 2 to 20°C sec^{-1}.

A few minerals exhibit what can be considered as single peaks and hence can be easily analyzed by conventional kinetic techniques. However, most mineral glow curves are rather broad, indicating the presence of more complex trap structures. Broad peaks are common for fine-grain pottery samples for which no mineral separation has been performed. Kinetic studies have been carried out on single peaks in calcite and quartz, which are well separated from any other glow peaks. Most crystalline calcite shows a well-defined peak occurring at 275°C when a heating rate of 5°C sec^{-1} is used (Figure 1A). Various experiments have shown this peak to obey first-order kinetics, have a trap depth of 1.75 $\pm$ 0.03 eV, and a preexponential factor of 4 $\times$ 10^{15} sec^{-1}.[20] From these values a mean life of 110 $\times$ 10^6

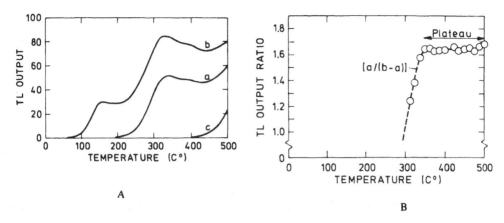

FIGURE 2. (A) Typical glow curves for fine grains extracted from pottery: (a) natural TL; (b) natural TL + TL induced by laboratory irradiation; (c) background "red-hot" glow. (B) Plateau test of TL storage stability over archaeological times; the value at each temperature equals a/(b − a) from (A). (After Aitken, M. J. and Fleming, S. J., *Topics in Radiation Dosimetry, Supplement 1*, Attix, F. H., Ed., Academic Press, New York, 1972, 1.)

years is predicted for an ambient temperature of 10°C, and hence the peak should be stable enough to be used for dating stalagmitic calcite formed in caves during the last million years.

Another peak that also exhibits first-order kinetic behavior is one seen in most quartz samples at 100°C when using a heating rate of 8°C sec^{-1} (Figure 1B). It does not appear in the natural TL since the electron traps are emptied in a few hours at ambient temperature.[21] It is a very important peak, however, since its sensitivity depends upon its radiation and thermal history and is used in the predose dating technique (Section IV.A.3). The two higher temperature peaks in quartz have also been subjected to kinetic analysis in order to ensure that they are sufficiently stable for dating. Fleming[21] found that the upper peak was certainly stable but that the lower peak, occurring at 325°C for a heating rate of 20°C sec^{-1}, should have decayed relative to the 375°C peak in a 2000-year-old pottery sample. However this was not the case, implying an underestimate of the lifetime of the 325°C peak which had been obtained by kinetic studies.[22] Later studies by Wintle[23] showed that the 1.05 ± 0.03 eV trap depth obtained by initial rise studies was much lower than that obtained by two other methods. Isothermal decay gave 1.7 ± 0.1 eV, and Hoogenstraaten's method,[24] which uses the relationship between peak temperature and heating rate, gave 1.69 ± 0.02 eV. Further experiments confirmed that the failure of the initial rise method was due to thermal quenching of the luminescence centers used by the 325°C peak.

Second-order kinetic behavior has been reported for several glow peaks in the feldspar mineral albite.[25] Theoretical aspects of the effect of second-order kinetics in geological materials have been presented by Levy.[26,27]

In practice it is usually not possible to apply simple kinetic analysis because most materials have overlapping peaks. In order to cope with this problem Aitken[28] developed an "ordinate ratio test", more commonly known as the "plateau test". Its simplest form is shown schematically in Figure 2, where the ratio of the natural TL to that induced by subsequent laboratory irradiation is compared. Such a plot enables one to say that in this sample there has been no loss of TL above 300°C. To get around problems due to sensitivity changes, a slightly different plateau test was developed in which $N/[(N+\beta) - N]$ is plotted vs. temperature, where N is the natural TL and $N + \beta$ is that from an identical sample which has been given an additional β dose.[22] A more stringent test of the suitability of the sample for dating is the existence of a plateau in the plot of calculated TL age vs. temperature (see, e.g., Gunn and Murray[29] and Mejdahl and Winther-Nielsen[30]).

The plateau test has also been employed to show that there is no spurious signal included in the natural TL, although spurious TL has to a large extent been eliminated by evacuating the TL oven and then filling it with an inert gas, such as nitrogen or argon, before measuring

Table 1

RADIOACTIVE DECAY SCHEMES OF THORIUM, URANIUM, AND POTASSIUM[11]

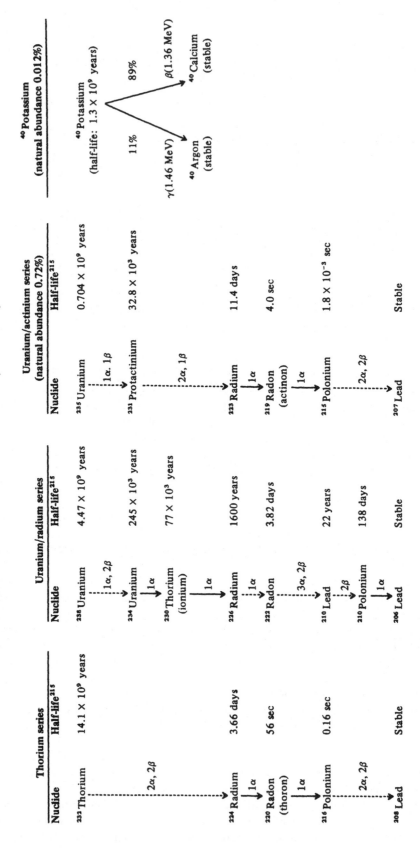

Thorium series

Nuclide	Half-life[215]
232 Thorium	14.1 × 10⁹ years
(2α, 2β)	
224 Radium	3.66 days
(1α)	
220 Radon (thoron)	56 sec
(1α)	
216 Polonium	0.16 sec
(2α, 2β)	
208 Lead	Stable

Uranium/radium series

Nuclide	Half-life[215]
238 Uranium	4.47 × 10⁹ years
(1α, 2β)	
234 Uranium	245 × 10³ years
(1α)	
230 Thorium (ionium)	77 × 10³ years
(1α)	
226 Radium	1600 years
(1α)	
222 Radon	3.82 days
(3α, 2β)	
210 Lead	22 years
(2β)	
210 Polonium	138 days
(1α)	
206 Lead	Stable

Uranium/actinium series (natural abundance 0.72%)

Nuclide	Half-life[215]
235 Uranium	0.704 × 10⁹ years
(1α, 1β)	
231 Protactinium	32.8 × 10³ years
(2α, 1β)	
223 Radium	11.4 days
(1α)	
219 Radon (actinon)	4.0 sec
(1α)	
215 Polonium	1.8 × 10⁻³ sec
(2α, 2β)	
207 Lead	Stable

40 Potassium (natural abundance 0.012%)

40 Potassium (half-life: 1.3 × 10⁹ years)

- 89% β(1.36 MeV) → 40 Calcium (stable)
- 11% γ(1.46 MeV) → 40 Argon (stable)

Table 2
DOSE RATE CONVERSION VALUES (μGy/a) FOR THE NATURALLY OCCURRING RADIONUCLIDES[38]

Radionuclide	Concentration	α	β	γ
Thorium series	1 ppm ^{232}Th			
No thoron loss		738	28.6	51.4
100% thoron loss		309	10.3	20.8
Uranium series	1 ppm ^{238}U			
No radon loss		2783	146.2	114.8
100% radon loss		1262	60.9	5.6
Natural potassium	1% K$_2$O		689.3	206.9
	1% K		830.3	249.2
Natural rubidium	100 ppm Rb		46.4	

Note: A more recent value of potassium γ doses are 200 and 241 μGy/a for 1% K$_2$O and 1% K, respectively.[217]

the TL.[31] A "good plateau" has also been used to show that a sample has been fired to a sufficiently high temperature and that the material has not experienced anomalous fading.

Anomalous fading is the name given to a loss of TL that cannot be explained in terms of the thermal decay predicted by kinetic theory.[32] The phenomenon was first discovered in samples of feldspar extracted from lava flow and was found to be the reason why the ages obtained by the TL method were always too young by up to an order of magnitude.[33] It was also found to occur in lunar samples[34] and in zircon and fluorapatite grains.

The ability to use the natural TL of a particular mineral for dating an event in its archaeological history relies upon its having been formed at that time, or that any memory of its earlier radiation history was destroyed at that time. Examples of formation are the precipitation of calcite by ground water in a limestone cave to form stalagmites and the formation of volcanic products, such as lava and ash. The most obvious zeroing of the TL signal is caused by heating and for this reason much effort has been expended on the dating of pottery and burnt stones from archaeological sites. Sunlight is also efficient at zeroing the TL signal in some minerals, particularly feldspars. This has recently led to the use of TL for dating recent sedimentary deposits that are known to have been exposed to sunlight prior to their deposition, e.g., loess and deep sea sediments. The zeroing ability of other processes such as pressure, grinding, and chemical changes has not been fully explored.

B. Natural Radiation Sources

The major sources of radiation in the natural environment are the uranium and thorium decay chains and the common radioactive isotope of potassium, ^{40}K; these decay chains are shown schematically in Table 1. A smaller contribution comes from cosmic radiation and an even lesser one from the decay of the common radioactive isotope of rubidium, ^{87}Rb.[35]

When dose rates are estimated from concentration measurements, the factors for converting concentrations of uranium, thorium, and potassium to α, β, and γ dose rates must be accurately known. In principle the calculation of these factors is straightforward, but because of the large number of transitions in the uranium and thorium series, the task is formidable. The values used in this chapter are those calculated by Bell[36] with later revisions.[37,38] In the calculations Auger and internal conversion electrons were included in the beta-ray contribution and the X-rays in the gamma-ray contribution. The 1979 values are listed in Table 2.

Typical concentration values of the radionuclides in Scandinavian pottery or soil are 10 ppm thorium, 3 ppm uranium, 2.5% potassium, and 200 ppm rubidium. The corresponding annual dose contributions are given in Table 3.

Table 3
INFINITE-MATRIX DOSE RATES (μGy a^{-1}) IN TYPICAL POTTERY AND SOIL
(10 ppm Th, 3 ppm U, 2.5% K, AND 200 ppm Rb)

Radionuclide	α	β	γ	Total	Effective total
[232]Thorium series					
Radionuclides before thoron	3,090	103	208		
	3,380	286	514	8,180	1,907
Thoron and later products	4,290	183	306		
[238]Uranium series					
Radionuclides before radon	3,786	183	17		
	8,349	439	344	9,132	2,035
Radon and later products	4,563	256	327		
[40]Potassium	0	2,076	623	2,699	2,699
[87]Rubidium	0	93	0	93	93
Total	15,729	2,894	1,481	20,104	—
Effective[a] dose rates	2,359	2,894	1,481	—	6,734
Relative contribution (%)	35	43	22		

[a]   Assuming an a-value of 0.15.

Recalculated from Aitken (Table 3.2)[11] using the conversion values in Table 2.

[40]K and [87]Rb give rise to β- and γ-radiation alone, but the uranium and thorium decay chains also emit α-radiation and this leads to very complex dosimetric conditions in mixed mineral samples such as pottery. Until the mid 1960s it was thought that α particles were almost totally responsible for the TL signal, since they deposited more energy per unit mass in the mineral crystals than did β- or γ-radiation. However, Zimmerman[39] showed that α particles are much less efficient at producing TL than β- and γ-radiation, in agreement with studies made on other dosimetry materials. Typical α-efficiency factors for the common minerals are in the range of 0.08 to 0.3.

The evaluation of the dose contribution from the uranium and thorium decay chains is less straightforward than that from [40]K for another reason: the decay chains may not be in secular equilibrium. Disequilibrium is brought about by the different physical and chemical properties of the various elements in the decay chain. For example, part of the way down the two decay chains are two gaseous elements: radon ([222]Rn, $t_{1/2}$ = 3.83 days) and thoron ([220]Rn, $t_{1/2}$ = 55 sec). The former is able to escape from certain types of minerals and from pottery if it is sufficiently porous.[40] If there has been radon escape from the pottery while it was buried, then the rest of the decay chain will not be present and its contribution to the dose rate will be lost.

Greater deviation from equilibrium occurs in two geological environments which have been the subject of TL studies. In the case of calcite formed by precipitation in limestone caves, [238]U and [234]U are taken into the $CaCO_3$ lattice. However, no [230]Th is incorporated because thorium is relatively insoluble and therefore undissolved by the circulating ground water. Decay of [234]U will produce [230]Th in the calcite with time, and indeed the return to equilibrium, as monitored by α-spectrometric techniques, gives the age of the calcite (the uranium series disequilibrium method). If TL dating is to be applied to the calcite, a time-dependent dose rate must be used.

The reverse situation holds on the ocean floor where [230]Th is precipitated in excess following its production in the ocean from the decay of [234]U. Once again a time-dependent dose rate must be applied, although this time the dose rate decreases with time.

Cosmic radiation typically contributes the order of 5% of the total dose rate in the midlatitudes for typical archaeological excavations. Lower values are applied when samples are from greater depths, e.g., in deep cave systems or at the bottom of the ocean.

C. Determination of Age

Nearly all TL dating techniques rely on the increase of the TL signal with time after the sample has had its earlier TL signal zeroed. The increase is due to the exposure of the mineral to the radiation produced by the decay of natural radioactive elements in its environment. Hence, the amount of TL is proportional to the radiation dose received by the mineral, provided that the response is linear. The linearity of the response is checked by laboratory irradiation of some samples prior to measurement of their TL. This enables the TL sensitivity to be determined. Hence, combining the natural TL measurements and TL sensitivity with separate measurements of the radioactivity of the sample enables us to write an age equation:

$$\text{Age} = \frac{\text{Total radiation dose}}{\text{Radiation dose per year}} \qquad (1)$$

Total radiation doses are quoted in gray (Gy) and annual doses in Gy/a.

This simple age equation is modified in practice according to the particular material being studied. As mentioned earlier, pottery dosimetry is extremely complex. Grains with a diameter of less than about 10 μm will be affected by α- and β-radiation produced within the pottery and by γ-radiation from the surrounding soil. However, grains of quartz, which contain negligible radioactivity, will receive a progressively smaller dose from α-radiation as the grain size increases. The β component changes little with grain size up to about 100 μm. For larger grains, appropriate β attenuation factors will have to be applied. As a result of these facts, pottery dating concentrated initially on the study of the TL of two particular grain size fractions, polymineral fine grains with a diameter in the range of 2 to 8 μm and quartz inclusions with a diameter of 90 to 125 μm.

For 90 to 125-μm quartz grains that have had their outer surface removed by chemical etching, the contribution from α-radiation will be negligible. Hence the annual radiation dose D_A is given by

$$D_A = 0.95D_\beta + D_\gamma + D_c \qquad (2)$$

where D_β and D_γ represent the infinite-matrix β and γ dose rates calculated from the measured values of ^{40}K and the uranium and thorium contents. The factor 0.95 allows for the attenuation of the beta dose due to the finite size of these grains.[41] D_c is the cosmic dose rate.

More recently, the inclusion technique has been applied to quartz grains up to 0.5 mm and also to alkali feldspar grains in the range 0.1 to 10 mm. The appropriate attenuation and self-dose (from ^{40}K in the feldspar lattice) factors can be obtained from the tables computed by Mejdahl.[41]

For 2 to 8-μm fine grains the dose rate includes an additional contribution from alpha particles, although this contribution is reduced by a factor which allows for the lower production of TL by α-radiation relative to β- and γ-radiation.[39] Hence the annual dose rate was originally given by

$$D_A = kD_\alpha + D_\beta + D_\gamma + D_c \qquad (3)$$

where k is the efficiency factor obtained using a ^{210}Po alpha source for laboratory irradiation. In a recent study, Aitken and Bowman[42] proposed a new system for specifying alpha source strengths which enables them to write an age equation that does not depend upon the type of laboratory source used. The annual dose rate is given by

$$D_A = 1.78aC_\alpha + D_\beta + D_\gamma + D_c \qquad (4)$$

where a, the a-value, is an efficiency factor which is equal to the k-value when 3.7-MeV alpha particles are used and C_α is the alpha count rate/ks cm^2.

III. DETERMINATION OF DOSE RATE

A. Alpha Dosimetry

1. Alpha Counting

From the very first consideration of the use of TL for dating[5] it was realized that α-radiation from uranium and thorium decay chains plays an important role in the production of TL in naturally occurring minerals. At that time the lower efficiency of α-radiation had not been reported and it was assumed that α-radiation was the dominant cause of natural TL. The α-activity was determined using a scintillation technique very similar to that commonly used today.[43] A silver-activated zinc sulfide phosphor powder was sprinkled onto a 2.25-in. diameter circle of Scotch® tape to make a uniform screen to be positioned on a PM tube. Crushed rock was then placed on the screen and the pulses of light that were emitted when an alpha particle struck the screen were counted. The count rates obtained were compared with those from uranium and thorium ore standards.

Since those early days, the popularity of alpha counting has fluctuated as there has been further understanding of the uranium and thorium decay chains. The supreme advantage of the technique is its basic simplicity. It involves very little operator time and the equipment is electronically stable and relatively inexpensive. Over the last 10 years ZnS:Ag spread on a plastic substrate has been commercially available and such screens have been widely adopted. They have greater reproducibility and have lower background counts than those made with ZnS:Ag powder.[44,45]

A major advantage of using alpha counting for the determination of the alpha contribution to the annual dose rate is that, in most cases, the count rate can be put directly into the age equation, i.e., it is not necessary to go through calculations which involve U and Th concentrations and dose-rate conversion factors. (The exceptions to this are those samples which are not in secular equilibrium, e.g., deep-sea sediments.) It is unnecessary to know the Th/U ratio since it has been shown that the error in the alpha contribution is less than 1.7% when the ratio is in the 1.2 to 10.5 range.[46]

When alpha counting is used to assess the beta and gamma contributions from the uranium and thorium decay chains then either equal U and Th activities (i.e., equal alpha count rates) are assumed or some attempt to separate the components is made using the ''pairs'' technique. For this technique the alpha counting electronics are expanded so that when a count is registered, another circuit is switched on for a short time (typically 0.4 sec) and this registers any other count arriving within that time. Most of these counts will be due to the rapid pair of alpha transitions in the Th decay chain,

$$^{220}Rn \xrightarrow{\alpha} {}^{216}Po \xrightarrow{\alpha} {}^{212}Bi \qquad (5)$$
$$t_{1/2} = 55 \text{ sec} \quad t_{1/2} = 0.145 \text{ sec}$$

where $t_{1/2}$ is the half-life. Discussions of the principles of ''pair'' counting can be found in Turner et al.[47] and Giffin et al;[48] more recently it has been extended further to include an even faster pair of alpha transitions in the ^{235}U chain.[49] The main problem encountered with the ''pairs'' technique is the need for long counting times to obtain a statistically meaningful number of pair counts.

An additional, more important, problem connected with alpha counting is the effect of disequilibrium in the decay chains. For the well-defined situations occurring in stalagmites or ocean sediments, where ^{230}Th is initially totally absent or in excess, alpha counting can be used. However, much thought has been given to the effect of disequilibrium due to diffusion of gaseous elements in the decay chains, in particular ^{222}Rn. Emanation of radon from pottery has been given as the explanation for the increased count rates observed after samples have been stored in their alpha counting cells which are sealed with an O ring. Various techniques have been devised to overcome this problem, e.g., using a "gas cell" to measure the emanation from a particular shard.[50]

Pernicka and Wagner[51] compared α activities for crushed pottery obtained by neutron activation analysis and fission track counting with those obtained by alpha counting. They observed large discrepancies in the results, both with sealed and unsealed alpha counting cells and concluded that the predominant cause was overcounting due to radon emanation.

Murray,[52] using an unsealed cell, compared alpha counting measurements of soils and pottery with values predicted from high resolution gamma spectrometry measurements on the same samples. The alpha counts were in excess by as much as 20 and 40%, respectively, for the two groups of samples. Murray suggested that the problem could be overcome by putting the sample in a resin so that the air gaps are filled.

Prescott and Jensen[53] described a method of fusing the powdered sherd or soil into glass discs using lithium borate. The method seems to eliminate effectively the factors contributing to overcounting. As a result of a comparative study they concluded that alpha counting, in spite of its limitations, is the most convenient method for determining alpha activity.

When alpha counting is used to evaluate β and γ dose rates in samples in which disequilibrium is present, as in the case of ocean sediments,[49] calculation of the dose-rate conversion factors requires a detailed knowledge of the α-particle range. The α energies for the uranium and thorium decay chains have been tabulated by Bowman[54] and the average ranges in pottery of these α particles have been calculated by interpolation between the ranges for Ne and Na. Alpha particle ranges in other materials, e.g., zircon or calcite, can be calculated using these tabulated energies.

2. Fission Track Analysis

The induced fission track technique employing thermal neutrons has been used to map the U contents of pottery.[55] Most of the tracks observed in the mica detector appeared to be randomly distributed and gave concentrations of ~3 ppm, but occasionally a "star" of tracks was observed. The minerals giving rise to these "stars" were found to be zircon and apatite which are known to have high U concentrations. Malik et al.[56] have used a fission track method to obtain the spatial distribution of U in a variety of archaeological samples including pottery and flint. They compared the distribution with the pattern of the TL obtained by direct photography of the slices as they were heated after gamma-irradiation. Fission track maps have been used for studying the distribution of U in slices of calcite,[57] flint,[58] and granite.[59] The technique has also been used to look for "zones" of U in zircon grains.[60]

A double irradiation technique with thermal and fast neutrons has been employed for determining the U and Th contents of powdered pottery and soils.[61] It is interesting to note that for application to TL it may be sufficient to irradiate only once with fast neutrons since U undergoes fission with a probability of 4.5 times that of Th. Since the Th/U ratio is usually around 3 to 4, this means that the dose rates could be calculated from the total track density in the detector in a way that is analogous to alpha counting dose rate calculations.

3. Neutron Activation Analysis

The use of neutron activation analysis (NAA) has frequently been proposed as a method of determining U and Th contents by TL laboratories with easy access to a reactor.[51] Delayed neutron counting has been shown to be capable of rapid analysis of U in pottery samples.[62] The Th content can be observed with sufficient accuracy 1 week after irradiation, although slightly improved accuracy can be achieved after 3 to 4 weeks delay. A recent comparison of NAA and alpha counting[51] has shown that NAA has superior accuracy unless special precautions are taken to prevent overcounting.[53] One disadvantage of the NAA methods is that only the parent U and Th are obtained; hence greater errors will be introduced by this method than by alpha counting in situations where there is disequilibrium of the decay chains.

4. Gamma Spectrometry

Very little use has been made of gamma spectrometry for determining the U and Th contents of archaeological material, although it has been used extensively for geological samples. Meakins et al.[63] have shown that it is possible to obtain the U, Th, and K contents for a shard of about 100 g using a suitably shielded sodium iodide detector. The technique is subject to error if there is disequilibrium in the decay chains since it involves measurement of daughter isotopes. In particular, the gamma decay used for the uranium determination is that of ^{214}Bi, which is well down the decay chain and hence will be affected by any radon loss. Combining these measurements with NAA they were able to demonstrate the existence of disequilibrium, but could not pinpoint its source.

The development of solid-state detectors has opened up the possibility of high resolution gamma spectrometry which can locate the source of the disequilibrium. Nielsen[64] used a Ge(Li) detector in the field for investigating disequilibrium in Danish soils. Murray and Aitken[65] used an intrinsic Ge well-type detector in a special low background housing to study the disequilibrium in a variety of shards and soils. However, in archaeological dating laboratories, high resolution gamma spectrometry is likely to remain a specialized tool for disequilibrium studies because of the expense and complexity of the equipment.

5. TL Efficiency Factor

As mentioned in an earlier section, the alpha contribution to the natural TL is reduced by a factor which allows for its lower TL production. Zimmerman[39] carried out a comprehensive study on several phosphor samples using fine grains (1 to 8 μm); irradiations were made with a ^{90}Sr/^{90}Y beta source and a ^{210}Po alpha source (3.7-MeV alpha particles). For these sources the alpha efficiency factors for quartz were found to be 0.10 for the 375°C peak and only 0.018 for the 110°C peak. In the second half of his paper Zimmerman considered the various causes of low alpha particle efficiency and concluded that the experimental evidence implied that it is a result of the spatial energy density around the center of the alpha track. In the region of high energy density the TL response is in saturation and hence less TL is obtained. See Volume II, Chapter 3, Section V.C for the modern track structure theory quantitative approach to this subject. At high doses the alpha and beta responses become similar and have the same saturation value. Zimmerman also found a relationship between the alpha efficiency and beta saturation doses for the various phosphors, which confirmed the hypothesis.

In nature, however, the TL induced by alpha radiation is not produced by monoenergetic 3.7-MeV alpha particles and it is necessary to sum the contributions from each of the alpha particles in the uranium and thorium decay series. The procedure is given in detail in Zimmerman's pioneer paper on the fine-grain technique.[66] In practice the efficiency factor known as the "k-value" is measured as the ratio of the TL sensitivities to known doses of α- and β-radiation, i.e.:

$$k = \frac{(\text{TL per Gy of } \alpha\text{-radiation})}{(\text{TL per Gy of } \beta\text{-radiation})} \tag{6}$$

This approach was used until 1975 when Aitken and Bowman[42] developed a more fundamental methodology. The main drawback of the k-value method is that since the ionization density increases as the alpha particle energy decreases, the TL effectiveness is strongly energy dependent. Hence the k-value has to be recalculated if a different laboratory alpha source is used. However, Zimmerman noticed that the TL per unit length of track for alpha particle energies ranging from 1 to 3.7 MeV was approximately independent of the energy.[39] This finding was confirmed and the study extended to higher energies by Aitken and Bowman[42] in experiments carried out using a Van der Graaf generator to produce particles up to 7.3 MeV. In the new system the alpha source strength is defined in terms of the rate at which track length is delivered to unit volume of the sample being irradiated. For convenience, a parameter, a, is introduced such that the annual alpha dose rate can be defined in a way similar to that used in the k-value method; this parameter is chosen so that for a pottery sample irradiated with 3.7-MeV alpha particles the a-value will be numerically identical with the k-value. The a value system has now been adopted by most TL dating laboratories.

6. Attenuation of Alpha Radiation

Another important aspect of alpha-radiation dosimetry is the attenuation of the alpha radiation. This is very important for dating pottery since the maximum range of the naturally occurring alpha emitters is about 45 μm in pottery and the emitting elements are not uniformly distributed. For example, pottery usually contains inclusions of quartz, which have very low U and Th contents, but these inclusions are surrounded by a clay matrix that contains U and Th. The outside part of a 100-μm grain, therefore, receives alpha radiation from these elements, but the inner core receives virtually no alpha dose. In the quartz-inclusion technique the outside of the grain is etched with HF acid.[67] This was originally performed to improve the transparency of the grains for TL measurement and to remove other minerals such as feldspars. However, removal of a 6-μm layer by treatment with concentrated HF for 40 min has the added advantage of removing the region irradiated most densely by alphas. The alpha contribution to the natural TL is thus reduced to an almost negligible level as shown by a plot of residual U and Th alpha dose vs. etching depth.[68] These residual components have recently been recalculated.[69]

Bell[70] has also computed the alpha attenuation factors for quartz grains ranging from 1 to 1000 μm in diameter. This should encourage people to use grain sizes other than the 1- to 8- and 90- to 125-μm sizes currently favored. Another point raised by this study is that even for fine grains, 1 to 8 μm, there is a 5% attenuation in the alpha dose contributed if the grains themselves contain negligible internal radioactivity. This effect is also important for the 4- to 11-μm grains used by Wintle and Huntley[71] for the TL dating of ocean sediments since a large proportion of the dose comes from ^{230}Th that is assumed to be held onto the grain surface.

A further reduction in the amount of alpha dose absorbed by mineral grains is brought about by the presence of water in the pores of the pottery. For U and Th alpha particles the energy absorption per gram in water is 50% higher than the energy absorption per gram of pottery. This leads to a true alpha dose rate

$$D_\alpha = \frac{D_{\alpha,dry}}{1 + \left(\dfrac{\text{sample weight } in \ situ}{\text{sample weight dry}} - 1\right) \times 1.50} \tag{7}$$

which is often rewritten

$$D_\alpha = \frac{D_{dry}}{1 + H_\alpha \Delta} \tag{8}$$

where $H_\alpha = 1.50$ and is the ratio of the specific stopping powers of water and pottery and

$$\Delta = \frac{\text{Weight of water } in \ situ}{\text{Weight of dry sample}} \qquad (9)$$

7. Alpha Dose from Inherent Radionuclides in Zircon, Calcite, Quartz, and Feldspar

The preceding discussion of the role of alpha radiation in TL dating has looked at the dose absorbed in a grain assuming that the source of the alpha activity is external to the grain. In this section we will consider a variety of different minerals and assess the alpha contribution from U and Th within the grain itself. Zircons, and to a lesser extent apatite grains, have a high U content — ~50 to 3000 ppm — and their presence in a pot can be detected using fission track analysis. Such high U values give rise to effective annual dose rates at the center of a grain ranging from 0.1 to 1.2 Gy a^{-1} [60] compared with typical alpha contributions to the fine-grain clay matrix of about 3 mGy a^{-1}. The advantages of using zircon grains for dating will be outlined in Section V.E.1 but here we will discuss the problems involved in measuring the annual dose rate. Since each grain requires its own U and Th measurements, induced fission track techniques are employed on each grain after all the TL measurements have been completed. The decay chains may be considered to be in secular equilibrium because of the geological age of such grains. The infinite alpha dose rate per part per million of each radionuclide must be recalculated for zircon, since the mean range (in millimeters) of an alpha particle in zircon is about half that in quartz. An added difficulty is the loss of alpha dose rate near the surface of the grain which must be allowed for in the dose rate calculation for each individual grain and the most appropriate correction factor applied, whether for a cube or sphere.

A more basic dosimetric problem besets zircon dating, however. Fission track maps often show the U concentration to be inhomogeneous and comparison of the maps with cathodoluminescence photographs implies anticorrelation of the U content and the luminescence sensitivity, i.e., regions that are more luminescent tend to be low in U and vice versa.[60] In this situation the natural dose determined by ignoring this effect will be too low. Similar problems of "zoning" are thought to upset the TL dating of some stalagmitic calcite.[72] In theory, if either the U content or the luminescence is uniform, then TL dating can be applied.

In the case of calcite there is yet another difficulty to overcome. When the stalagmite is formed, ^{238}U and ^{234}U are taken into the calcite lattice, but for geochemical reasons negligible ^{230}Th is incorporated. This gives rise to a time-dependent dose rate[72] since ^{230}Th is formed by the subsequent decay of ^{234}U in the calcite. The alpha contribution is often of far greater importance for stalagmites than for pottery for two main reasons. First, the a-values are considerably higher; Wintle found a-values ranging from 0.24 to 0.56 which may be compared with values of about 0.1 for most pottery and detrital sediments. Second, the gamma contribution is likely to have varied in the past due to changes in the surrounding cave environment.

Assuming that quartz inclusions contain negligible uranium and using alpha attenuation factors, Fleming[67] calculated the effective alpha dose rate for 100-μm inclusions which had had their surface removed by HF etching. He concluded that it was negligible. However recent studies have suggested that quartz grains may contain sufficient uranium to make as much as a 5% contribution to the total dose rate.[73] Also, other studies have shown that HF etching does not always proceed isotropically and that in some grains the etching creates tunnels into the center of the grain.[74] This would also affect the alpha particle contribution. Fortunately the low a-values for quartz prevent large errors in the calculation of the effective TL produced by alpha radiation. More recent HF etching experiments have been carried out on unfired quartz grains which had previously been annealed and then irradiated using a specially constructed multisource alpha irradiator.[75]

In the case of TL dating of pottery using alkali feldspars, the a-values are higher than for quartz but the internal dose rate due to beta-radiation from the ^{40}K decay makes the alpha contribution almost negligible.[30] Plagioclase feldspars, on the other hand, have a lower K content than alkali feldspars and even higher a-values. For 100-μm grains the effective alpha particle contribution may be as high as one third of the total dose rate.[76] The alpha dosimetry is not straightforward, however, since the feldspar inclusions also have a low U content, as shown by fission track mapping. In an attempt to simulate the natural alpha-particle flux, the grains were irradiated in a multisource irradiator.[75] Details of the dosimetry and results for feldspars have been presented by Guérin.[77] Plagioclase feldspar TL is now used routinely to study the development of volcanic activity in the Massif Central, France.

B. Beta Dosimetry

In most cases beta particles originating from radionuclides distributed in the samples contribute the predominant part of the radiation dose accumulated in archaeological and geological samples. For quartz and feldspar grains larger than 0.1 mm, the contribution can be as high as 60 to 80%, and in fine-grain dating about 50% of the TL emission results from the beta radiation because of the reduced efficiency of alpha radiation in producing TL. Consequently, considerable attention must be given to an accurate determination of the beta dose rate.

1. Sources of Beta Radiation

Beta radiation is produced by the decay of the naturally occurring radionuclides ^{238}U, ^{235}U, ^{232}Th, ^{40}K, and ^{87}Rb discussed in Section II.B. Because the average range of beta particles is short (1 to 2 mm in clay) compared with that of gamma radiation, the sources of beta radiation can be confined to the radionuclides occurring in the samples themselves. The exclusion of the environmental influence has the great advantage that the determination of beta dose rate involves only laboratory measurements. In the dating of archaeological samples, the effect of beta radiation from the surrounding soil is eliminated by removing the outer 1 to 2 mm of the samples. This precaution is not necessary with geological samples because these are usually derived from homogeneous layers with a uniform distribution of radionuclides.

In addition to the matrix beta dose rate discussed above, the grains used for dating in the inclusion method (quartz, feldspar, zircon) may experience beta radiation from inherent radionuclides. Using the fission track technique, Sutton and Zimmerman[73] found that some quartz grains had areas (less than 10 μm in diameter) that contained several hundred parts per million of uranium. They estimated that the inherent alpha radiation would contribute from 2 to 11% of the effective dose. Mejdahl et al.[78] found an alpha dose contribution of about 2% in quartz and feldspar grains. In zircon and apatite, which typically contain 100 and 30 ppm of uranium, respectively, the inherent dose is predominant. In small radioactive inclusions (~0.1 mm) the beta dose rate from inherent uranium and thorium is negligible in comparison with that from alpha radiation, while for larger grains (millimeters in size) it increases to about 50% of the alpha dose rate assuming an a-value of 0.15. Alkali feldspar may receive a substantial dose from potassium in the crystal lattice. In practice, alkali feldspars frequently contain 6 to 10% of potassium, the maximum being 16% in pure potassium feldspar ($KAlSi_3O_8$). As an example, a 0.5-mm grain containing 8% of potassium will experience a beta dose rate of 5.36 mGy a^{-1} [79] which is considerably more than that existing in a typical clay matrix (Table 3). The effect of beta radiation from rubidium has been estimated by Warren.[35] As the mean beta energy is only 0.104 MeV, the particle range is small (about 50 μm) and the effect virtually negligible in inclusion dating. However, in fine-grain dating ^{87}Rb contributes 1 to 2% of the dose, and Warren recommended that a potassium-to-rubidium ratio of 200:1 should be assumed if the rubidium concentration is unknown; this is equivalent to increasing the estimated beta dose rate from potassium by 2.9%.

In fine-grain dating the relevant beta dose rate is the matrix dose rate discussed above, corrected for attenuation by water if necessary (see Section III.B.3). The estimate of the beta dose rate experienced by larger grains is more complicated and proceeds in three steps: (1) estimate of the matrix dose rate as in fine-grain dating, (2) correction of this dose rate for attenuation of the radiation in traversing the grains, and (3) estimate of the contribution from inherent radioactivity in the grains.

2. Determination of the Matrix Beta Dose Rate
a. Introduction

There are two different approaches to the determination of the matrix or gross sample beta dose rate: (1) measurement of the contents of radionuclides in the sample and subsequent calculation of the dose rate assuming that energy absorbed equals energy released[38] and (2) direct determination of the dose rate by means of TL dosimetry.

b. Concentration Measurements

The methods used for concentration measurements are alpha counting, neutron activation analysis, fission track counting, and gamma spectrometry. Alpha counting is not recommended for determining beta dose rate unless the Th/U ratio of the sample is known.[80] The other techniques can be used with advantage provided that the U and Th series are in secular equilibrium. The state of equilibrium can be determined by means of high resolution gamma or alpha spectrometry. Gamma spectrometry yields the K content in addition to the U and Th contents, but the other methods require a separate determination of the K content by atomic absorption, flame photometry, or X-ray fluorescence.

c. Dose Rate Determination by TLD

The determination of the beta dose rate by means of sensitive TL phosphors such as CaF_2, $CaSO_4:Mn$, or $CaSO_4:Dy$ is an attractive method to give the present-day dose rate directly, irrespective of disequilibrium in the decay chains. Short-term disequilibrium caused mainly by the diffusion of radon poses some problems that will be discussed in a later section.

A basic problem is the achievement of a response of the TL phosphor that is independent of the ratio of the contents of thorium, uranium, and potassium. The dependence of the response on relative concentration values is caused by two factors:

1.   The need to shield the phosphor from alpha radiation from the sample introduces attenuation of the beta radiation and this is dependent on the beta spectrum.
2.   There is a significant gamma contribution from the sample, and the ratio of gamma-to-beta dose rate (in any infinite medium) is drastically different for the three contributors: 1.8:1 for the thorium series, 0.8:1 for the uranium series, and 0.3:1 for potassium (cf. Table 2).

The first TLD system for pottery dosimetry discussed by Aitken[28] consisted of pellets of $CaSO_4:Mn$ sandwiched between two fragments of pottery. The system inspired the development of a number of TLD methods; four of these are described below:

i. CaF₂ Embedded in the Matrix

This method was proposed by Fleming[81] and has been described in detail in Fleming.[68] The method involves mixing 100-μm grains of fluorite into a gram of powdered pottery and extracting them again by sieving after the requisite storage time. Alpha particle dosage is avoided by desensitizing the outer part of the fluorite grains by heating them in a moist atmosphere of 600°C before use. The method was used specifically for quartz-inclusion dating in which the outer layer of the grains was etched away by treatment with concentrated hydrofluoric acid. Because the absorption of the beta radiation in the fluorite grains replicates the absorption situation pertaining to the inner core of the quartz grains used for dating, the first part of the dependence on radionuclide ratios is eliminated; however, there remains some degree of dependence through the gamma contributions to the grains. The method has been used in dating studies by Fleming,[67] but despite its advantages it has not found routine application because of the technical difficulties involved.

The inclusion of the gamma contribution from the sample could be turned into an advantage by using the total sample for the beta measurement. In that case the total gamma self-dose would be included automatically with the beta dose. The gamma dose rate measured in the soil would then need to be reduced by an amount corresponding to the attenuation of the soil gamma in a mass of soil equal to that of the sample. This procedure would obviate the problem arising in gamma dosimetry if the radioactivity content in the sample differed from that in the soil.[82]

ii. Beta Dosimetry Based on Quartz and Feldspar Grains

McKerrell and Mejdahl[83] suggested the possibility of using the quartz and feldspar grains themselves as built-in phosphors for measurement of the beta dose rate in inclusion dating of pottery. The two phosphors have low temperature peaks with a sensitivity sufficient to measure doses in the mGy region. In quartz the 110°C peak can be activated to a high sensitivity by irradiating the sample with a large dose (~100 Gy) and subsequently heating it to 500°C. Feldspar has a sensitive peak at about 140°C.

Because the peaks are not stable at room temperature, the pottery samples must be stored at a low temperature (e.g., liquid nitrogen) during the irradiation period and the subsequent preparation of the grains for measurement (cleaning, etching, etc.) must be timed accurately.

Because dose and dose rate measurements are carried out on identical grains, it was thought unnecessary to etch away the alpha-irradiated outer layer of the grains; however, it turns out that the a-value for the 110°C peak in quartz is lower than that pertaining to the 375°C peak by as much as a factor of 10. Three advantages remain: (1) correction for attenuation of beta radiation is unnecessary, (2) self-dose from inherent radioactivity in the grains is automatically included, and (3) the gamma self-dose is included with the consequences discussed above. Technical difficulties have so far precluded routine application of the method, but the feasibility of using quartz grains for beta dosimetry has been demonstrated by Liritzis and Galloway.[84]

iii. Oxford β-TLD Unit

In a TLD unit for beta dosimetry developed at Oxford,[85] particular attention was given to the problem of dependence of the response on the ratio of radionuclide concentrations, and independence within 2% was achieved.

The solid, crushed, or powdered sample is held in a perspex® container (volume 9 cm³) whose bottom has a circular window of melinex® of a thickness (125 μm, 17.5 mg cm^{-2}) sufficient to stop alpha particles emitted by the sample. The unit is light tight but not gas tight, thus allowing the escape of radon and thoron. The TL phosphor, natural fluorite (MBLE type Super "S"), is fastened with silicone resin (Dow Corning MS 805) in a copper tray of 250-μm thickness and 30 mm in diameter. Figure 3 shows the unit with the copper tray in place underneath the melinex® screen. During the exposure period (about 2 weeks) each container is kept in a lead box with a wall thickness of 25 mm in order to provide intermatrix gamma shielding and reduce laboratory gamma radiation background. Background contribution was estimated by means of units containing granular quartz. The dosimeters are read out in a special glow oven at a linear rate of 5°C sec^{-1}. The heat source is a 100-W projector lamp. Measurements are in terms of peak heights rather than area. The dose is estimated by comparing the response with that induced by known doses from a ^{90}Sr/^{90}Y beta source. The infinite-matrix dose is obtained by multiplying the phosphor dose by

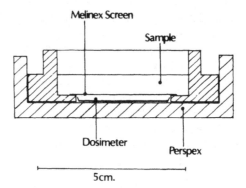

FIGURE 3. Oxford β-TLD unit. The pottery powder is held in an unsealed container whose bottom is a 125-μm sheet of melinex® plastic. The TL phosphor in the tray is natural calcium fluoride. (From Bailiff, I. K. and Aitken, M. J., *Nucl. Instrum. Methods*, 173, 423, 1980. With permission of the North Holland Publishing Company, Amsterdam.)

5.49, a factor derived from test experiments using matrices of known radioactivity content. The dose rate at the position of the phosphor from the sample composition considered in Table 3 is thus 1.7 μGy/day. The materials used for the test experiments included U.S. A.E.C. (New Brunswick) radioactive sands (diluted monazite and diluted pitchblende) and KCl. The experiments showed that the response of the β-TLD unit is independent of the matrix material to within about 2% and hence of any ratio occurring in archaeological and geological samples. Calculations indicate that the independence rests on two compensating effects: progressing from thorium to potassium the beta particles become more penetrating, but the resulting increase in beta dose rate is compensated by a diminution of the gamma contribution to the dosimeter.

Samples as small as 6 g can be measured with the unit, but with a slightly increased dependence of response on radionuclide concentration ratios. A smaller unit, designed along similar lines,[86] allows measurement of samples down to 1.5 g with a ±5% dependence of response on concentration ratios.

iv. Risø β-TLD System

The Risø β-TLD system is similar to the Oxford system but differs in certain details.[87,88] The powdered samples contained in polyethylene bags are heat sealed in tubes of black polyethylene foil. Normally, 30-g samples are used, but accurate measurements can be made on samples as small as 5 g. Pottery samples are usually dried before sealing, whereas burnt stones and sediment samples are sealed with their natural water content. The total shielding provided by the bag and foil is 24 mg cm^{-2}. The sealing prevents the escape of thoron and radon, and as a further precaution, the sealed samples are kept for 4 weeks before the beginning of the measurement to allow build-up of radon to equilibrium.

The dosimeter unit is a metal planchet (0.30 × 40 × 60 mm) made of kanthal. Encapsulated in silicone resin (Dow Corning MS 805) is 400 mg of $CaSO_4$:Mn fastened in a circular depression in the planchet (diameter 28 mm); the phosphor thickness is thus 65 mg cm^{-2}. The dosimeters are attached to the surface of the samples, one on either side (Figure 4). For small samples the planchet dosimeter is replaced by discs of teflon® (5 mm in diameter × 0.4 mm) impregnated with $CaSO_4$:Mn.

Because of the rapid fading of the $CaSO_4$:Mn peak at room temperature, samples and dosimeters are kept in a freezer at −25°C during the irradiation period, usually 1 to 2 weeks. For measurement, the dosimeters are heated at a rate of about 15°C sec^{-1} by means of an

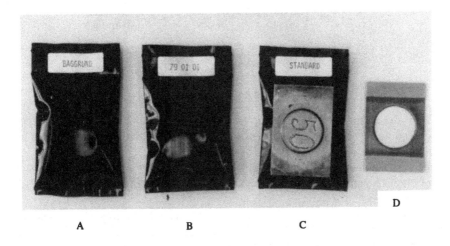

FIGURE 4. Risø β-TLD system. The pottery powder is sealed in polyethylene bags, thickness 24 mg/cm². (A) Nonradioactive sample. (B) Archaeological sample. (C) Standard clay with a dosimeter in place. (D) Dosimeter.

electric current passed through the planchet; the reading is expressed in terms of peak height. The apparatus is equipped with a sample changer that can accommodate 12 samples.[89] The teflon® discs are read out in the hot nitrogen reader designed by Bøtter-Jensen.[90] The dose recorded by the dosimeters is about 20% of the infinite-matrix dose.

The samples are stored in sets of six for dose rate measurement: two archaeological or geological samples, two standard samples made of a commercial pottery clay, and two nonradioactive samples (CaSO$_4$,2H$_2$O) for measurement of gamma-background radiation. The internal dose rate in the unknown samples is obtained by comparing the TL signals from the dosimeters with those induced by the standard samples after subtracting the gamma-induced signal. The internal beta dose rate in the standard samples was found by comparing them with a primary standard made by mixing potassium carbonate and uranium and thorium ore (U.S. A.E.C., New Brunswick Laboratories) in known proportions with a nonradioactive matrix (CaSO$_4$). Water-free CaSO$_4$ was selected because its stopping power is close to that of dry clay.

The response of the dosimeter is not independent of the ratio of U, Th, and K concentrations; therefore the system is directly applicable only to samples with U, Th, and K concentration ratios approximately equal to those in the standard samples for which the relative dose rate contributions of Th:U:K are 11:8:71. A correction is required for samples having a potassium contribution below 50%; it is thus necessary to measure separately the potassium content of the gross sample. The dependence of the response on the Th:U:K ratio is of little consequence in routine work because the majority of samples dated (mainly from Scandinavia) have potassium contributions exceeding 50% of the beta dose rate.

3. Attenuation by Water

Archaeological samples buried in the ground as well as those taken from geological deposits usually contain a certain percentage of water which attenuates the beta radiation. A correction for water attenuation is therefore necessary if the beta measurement has been carried out on a dry sample. The relation between the corrected dose rate R$_\beta$ and the dose rate R$_{\beta,dry}$ measured on the dry sample is given by the following equation.[66,91]

$$R_\beta = \frac{R_{\beta,dry}}{1 + \left(\dfrac{\text{sample weight } in \ situ}{\text{sample weight dry}} - 1\right)\dfrac{S/\rho(\text{water})}{S/\rho(\text{sample})}} \tag{10}$$

with

$$\frac{S/\rho(\text{water})}{S/\rho(\text{sample})} = 1.25 \tag{11}$$

where S/ρ is the stopping power. The correction may amount to 20% for pottery, depending on firing conditions, and up to 50% for sediments. The correction for burnt stones is usually very small, 2 to 3%.

Zimmerman[91] obtained the stopping power ratio of 1.25 by assuming that the stopping power of clay is equal to that of aluminum. Using the stopping power for dry clay, a ratio of 1.23 is obtained.[87]

In order to apply the equation it is necessary to estimate the average water content of the sherd during burial. Almost complete saturation has been found for ceramics from temperate geographical regions. Mejdahl and Winther-Nielsen[92] have shown that shards buried at least 50 cm below the present ground surface rapidly acquire a water content close to saturation, which is maintained throughout the year. For ceramics from warmer countries an average water content during burial must be estimated by combining the *in situ* water content and the saturation water content measured in the laboratory with environmental information such as annual rainfall and ground water level. If no other information is available, one can assume that the water content during burial is $\frac{1}{2} \pm \frac{1}{2}$ of saturation.

For sediments it is often more difficult to estimate the past water content since samples taken from exposed faces no longer retain their original water content. In this case an estimate of the original content must be made using a fraction of the saturation water content. Some information concerning this fraction is often available from hydrological studies on similar sediments; e.g., Audric and Bouquier[93] have measured seasonal changes in water content for loesses in Northern France.

4. Correction for Attenuation by the Grains

The beta dose rate resulting from correction for water attenuation is directly applicable in fine-grain dating, but in inclusion dating a further correction is necessary because of the attenuation of the radiation in penetrating the grains. Mejdahl[41] has calculated correction factors for attenuation by spherical quartz grains with diameters in the 0.005- to 10.0-mm range. The calculation was based on absorbed-dose distributions for point-isotropic beta sources as listed by Berger[94,95] combined with the scaling procedure developed by Cross.[96] This calculation included beta particles, internal conversion electrons, and Auger electrons emitted by ^{40}K and the ^{232}Th and ^{238}U series. The values are also valid for feldspars, which have stopping powers close to those of quartz (about 1% lower).

The resulting attenuation depends on the relative abundance of the radionuclides. Figure 5 gives the relative attenuation for the composition assumed earlier. The curve shown is valid for unetched grains. If grains are etched by concentrated HF, a further small correction is required, amounting to a maximum of 3% for removal of a layer of 8 μm from a 100-μm grain (etched for 60 min), and is negligible for larger grain sizes.

5. Beta Dose Rate to Alkali Feldspars from Inherent Potassium

Potassium in the crystal lattice may contribute significantly to the beta dose rate in alkali feldspar grains, in particular for grain sizes exceeding 1 mm. Absorbed beta dose fractions in alkali feldspar have been calculated by Mejdahl.[41] Annual beta dose to feldspar grains containing 1% K has been plotted in Figure 6 as a function of grain size. A potassium

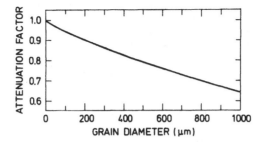

FIGURE 5. Attenuation of external beta radiation in quartz or feldspar grains as a function of grain size. Assumed radioactivity content of sample 10 ppm Th, 3 ppm U, and 2.5% K. Attenuation values taken from Mejdahl.[41]

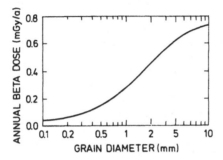

FIGURE 6. Average annual beta dose in alkali feldspars from inherent potassium as a function of grain size. Assumed potassium content 1% K. (From Mejdahl, V., *PACT*, 9, 351, 1983. With permission.)

content of 10% is not uncommon and the contribution from inherent potassium will thus be the dominant component for grain sizes exceeding 1 mm. A high contribution from inherent potassium has the advantage that the effect of uncertainties in the measurement of external beta and gamma radiation, e.g., from disequilibrium in the uranium series, will be very reduced.

6. Effect of Disequilibrium

Disequilibrium in the uranium series may be short term, arising from the escape of radon from the ceramics during burial, or long term if the clay from which the ceramics was made was deposited with a deficit or surplus of ^{238}U, ^{234}U, or ^{226}Ra relative to ^{230}Th.

The effect of radon escape depends on the relative contribution of the uranium beta dose rate. In the typical pottery considered earlier the uranium beta contribution is about 15% of the total beta contribution. The total error in the beta dose rate that would result if all the radon escaped from the ceramics during burial would thus be about 7%. Examples are known of shards having a very large emanation of this gas,[97] but, in general, radon escape is not a serious problem in inclusion dating of shards for which potassium contributes 60% or more of the beta dose. Care must be taken, however, in the case of samples where the uranium and thorium contributions predominate.

Deviation from equilibrium can be measured by means of high-resolution gamma spectrometry as described by Carriveau and Harbottle[98] and Murray and Aitken.[65]

Long-term disequilibrium in the uranium series arises as a result of the water solubility

Table 4
EFFECT OF
DISEQUILIBRIUM IN THE
URANIUM SERIES

Time (a)	Beta dose rate (μGy a^{-1})
0	53.7
10,000	61.5
30,000	74.3
100,000	107.1
300,000	136.2
1,000,000	140.4

Note: Build-up of beta dose rate with time when ^{238}U and ^{234}U have been deposited without their daughters as in the case of stalagmite. Values given are for 1 ppm uranium and ^{234}U/^{238}U = 1.15 at t = 0. Conversion factors are from Bell.[36]

Table 5
SOURCES OF ERROR AND
ESTIMATED UNCERTAINTIES
(1σ) ASSOCIATED WITH THE
RISØ TLD SYSTEM FOR
MEASUREMENT OF BETA DOSE
RATE IN POTTERY

Source of error	Uncertainty (%)
Measurement of TL	2.5
Dose rate of standard	3.0
Water content correction	1.5
Attenuation by grains	1.0
Radon loss	2.0
Total uncertainty	4.7

of uranium and radium compounds. Using a combination of techniques, Meakins et al.[63] described the effect of a surplus of radium in a shard, 3600 years old, from Jericho. The estimated present dose rate ($\alpha + \beta$) from the uranium chain was 1.08 mGy a^{-1}, whereas the dose rate 3600 years ago was 2.50 mGy a^{-1}. As contact with ground water could be excluded, they concluded that the disequilibrium existed in the clay from which the shard was made.

As another example let us consider the beta dose rate inside a stalagmite. As mentioned in Section III.A.7, ^{238}U and ^{234}U are taken into the calcite lattice when the stalagmite is formed. ^{230}Th produced by the decay of ^{234}U increases with time towards its equilibrium level and hence the annual dose rate is time dependent.[72] This increase is shown in Table 4 for a sample with an initial uranium content of 1 ppm and ^{234}U/^{238}U ratio of 1.15. This activity ratio is typical of stalagmitic calcite and is caused by fractionation of the uranium isotopes during weathering.

The effect discussed above would result in an underestimate of the age. An overestimate would result if the sediment had a surplus of ^{230}Th, a very real possibility in waterlaid sediments because of the low solubility of thorium compounds in water. The effect is utilized in the ionium method of dating deep-sea sediments.[99]

7. Estimated Uncertainty in Beta Dose Rate Determination

Mejdahl[88] estimated the total uncertainty in the determination of beta dose rate in pottery, as shown in Table 5. The estimate is valid for grains with no inherent radioactivity embedded in pottery in which potassium is the major source of matrix beta radiation. The uncertainty would be larger for samples where disequilibrium effects were significant. For alkali feldspar grains in which inherent potassium contributed a sizeable part of the beta dose, the uncertainty of the total beta dose rate would be appreciably smaller (about 3% in the case of equal contributions from the two sources).

An additional error arises if a significant exchange of radionuclides between the sample and the surrounding soil has taken place during burial. Hedges and McLellan[100] have carried out an investigation of the cation-exchange capacity of fired clay and ceramics. They conclude that an appreciable cation-exchange capacity does exist and predict that perhaps 10% or more of the uranium and thorium contents is potentially exchangeable. Also, a significant portion of the K^+ ions would be subject to replacement depending on the ground water composition. Murray[101] also measured the ion-exchange capacity, and to measure the changes in shard activities that occur as a result of such experiments, used high resolution gamma spectrometry in addition.

C. Gamma- and Cosmic-Ray Dosimetry

The dose rate from environmental gamma radiation can be determined either by calculations following measurements of the content of radionuclides in a soil sample taken at the site or by direct measurements at the site carried out by means of TL dosimetry or scintillation counting. The former procedure requires a separate estimate of the cosmic-ray dose rate, whereas the cosmic-ray contribution is automatically included in the results from on-site measurements.

1. Cosmic Radiation

The long-term variation in cosmic-ray intensity, which has such far-reaching consequences for radiocarbon dating,[102,103] does not affect TL dating because (1) normally, the relative contribution of cosmic radiation to the total dose is small (5 to 6% in quartz-inclusion dating and less in fine-grain dating) and (2) the important factor is the average intensity over the period of interest, not the instantaneous, initial value as in radiocarbon dating. However, in some circumstances, e.g., at very high altitudes or on coral island sites, the cosmic-ray dose may become a more significant fraction of the total dose.

Only the hard component of the cosmic radiation will be effective because the objects and materials of interest will generally be buried at depths ranging from 30 cm to several meters.

Prescott and Stephan[104] have given a detailed discussion of the contribution of cosmic radiation to the dose in TL dating. From an evaluation of published data they derived a dose depth curve valid for latitudes higher than 35°. The curve was in good agreement with experimental values obtained by Aitken.[105] Values at depths of 100 and 200 g cm^{-2} in standard rock (atomic number 11 and atomic weight 22) were 0.185 and 0.150 mGy a^{-1}, respectively. At greater depths the dose rate decreases at a rate of about 8.5%/100 g cm^{-2}. The latitude effect at lower latitudes results in a decrease of the dose rate by about 10% from 35° to the equator. An expression for the altitude effect was derived; for latitudes higher than 40° this becomes:

$$I(h) = 0.185 (0.25 + 0.75 \exp (h/4.1)) \tag{12}$$

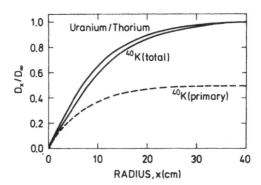

FIGURE 7. Gamma dose D_x absorbed at the center of a sphere of radioactive soil (density 2.7 g/cm³) as a function of the radius x of the sphere, relative to the dose D_∞ from a sphere of infinite dimensions. (From Fleming, S. J., *Thermoluminescence Techniques in Archaeology*, Clarendon Press, Oxford, 1979. With permission.)

I is the dose rate in mGy a⁻¹ to an object buried under 100 g cm⁻² of standard rock and h is the altitude in km. According to the equation the intensity at an altitude of 3.5 km will be twice that at sea level.

2. Gamma Radiation

The environmental gamma radiation to which a sample buried in soil is exposed originates from radionuclides in a sphere of about 30-cm radius surrounding the sample. Figure 7, taken from Fleming,[68] shows the absorbed gamma dose at the center of a sphere of homogeneous soil as a function of the sphere radius. The figure shows that 50% of the dose comes from a sphere with a radius of about 8 cm and 80% from one of 15-cm radius.

Typical gamma dose contributions from uranium, thorium, and potassium are given in Table 3.

3. Gamma Dosimetry from Concentration Measurements

Concentration measurements are made on the basis of 1 to 2 kg soil taken at the site and sealed immediately to preserve the natural humidity of the sample. The methods of measurement are those already discussed in the section on alpha dosimetry. The most reliable of these for gamma dosimetry would be gamma spectrometry because a large and consequently more representative sample can be measured in a sealed condition. The state of short-term equilibrium can be estimated by measuring the sample immediately after sealing and again after storage for a period sufficient to allow build-up of radon to equilibrium.

Except when a high resolution gamma spectrometer is used, gamma dosimetry by means of concentration measurements is particularly sensitive to disequilibrium effects since 95% of the gamma dose in the uranium chain is post-radon. In typical soil the uranium chain (in equilibrium) contributes about 25% of the gamma dose (Table 3).

4. Gamma Dosimetry by Means of TLD

a. TL Phosphors

The desirable properties of a TL phosphor for environmental measurements include

1.   Lower limit of measurement: 0.01 to 0.05 mGy
2.   Negligible fading over 1 to 2 years
3.   Easy to obtain or prepare

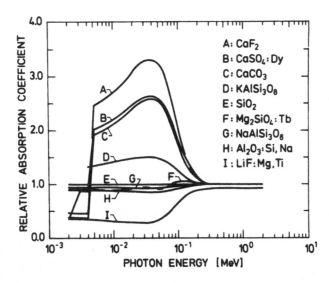

FIGURE 8. Variation of energy absorption coefficient (μ/ρ)$_{tot,en}$ relative to that for quartz for a number of TL phosphors. (From Mejdahl, V., *PACT*, 2, 70, 1978. With permission.)

4. Reusable
5. Simple annealing procedure
6. Negligible self-dose
7. Energy dependence of response equal to that of the minerals used for dating (quartz, feldspar, calcite)

The energy dependence of response for a number of TL phosphors and minerals is compared in Figure 8, taken from Mejdahl.[106] The graphs shown are energy absorption coefficients (μ/ρ)$_{tot,en}$ relative to quartz as functions of photon energy in the range of 0.001 to 2.0 MeV.

The following features are apparent from Figure 8:

1. CaF_2 and $CaSO_4$:Dy, two phosphors used in TL dating,[105,107] have a considerable overresponse relative to quartz if used unshielded.
2. $CaSO_4$:Dy unshielded would be ideal for estimating the dose to calcite.
3. Mg_2SiO_4:Tb and Al_2O_3:Si,Na represent very well the response of quartz and $NaAlSi_3O_8$.
4. LiF:Mg,Ti, which is used extensively in radiation protection measurements because of its tissue equivalence, will underestimate the dose to quartz.
5. The response of $KAlSi_3O_8$ is somewhat larger than that of quartz. The response of $Ca(AlSi_3O_8)_2$ (not shown) falls between that of quartz and $KAlSi_3O_8$.

Practical dosimetry systems based on CaF_2 (Oxford) and $CaSO_4$:Dy (Risø) have been developed; these will be discussed below. In both systems dosimeters are buried in the soil at the sites for periods of several months to 1 year in a situation that closely resembles that of the sample to be dated.

b. Oxford TLD System for Gamma Dosimetry

The gamma TLD system used at Oxford has been described in detail by Murray et al.[108] The TL phosphor is natural CaF_2 ("Super S" fluorite supplied by MBLE Brussels). This phosphor has good TL sensitivity and negligible fading. In order to reduce its overresponse relative to quartz the phosphor is contained in a copper capsule during the irradiation period. The dimensions of the capsule are wall thickness — 1.5 mm; inner diameter — 3 mm; and length — 15 mm. The capsules are sealed by hard soldering and hence can be annealed (at 500°C) in a portable oven before being buried at the site.

For measurement of the accumulated dose the height of peak III of the glow curve is compared with that induced by a subsequent irradiation of the phosphor with a laboratory ^{90}Sr/^{90}Y beta source that has been calibrated against a ^{137}Cs gamma source.

As the response of the CaF_2 dosimeter may not equal that of quartz exactly, a calibration experiment is required in which the dosimeter is irradiated together with quartz in an irradiation facility having a gamma spectrum similar to that in soil. A large concrete block (1 m^3) doped with 15 kg of uranium ore (uranium concentration 53.1%) was used for this purpose. The measured dose rate at the center of the block was 0.331 Gy a^{-1}. Obsidian rather than quartz was used for the experiment because of its higher TL sensitivity. The gamma absorption characteristics of obsidian are virtually identical to those of quartz. As a result of the experiment it was found that CaF_2 shielded by 1.5 mm copper underestimated the dose to obsidian by about 8%, whereas the response of CaF_2 shielded by 1.5 mm Al was almost identical with that of obsidian. However, copper capsules were preferred because of the significant levels of radioactive impurities found in aluminum.

c. Risø TLD System for Gamma Dosimetry

The system, described by Mejdahl,[106,107,109] is based on $CaSO_4$:Dy made at Risø. The phosphor is contained in polyethylene capsules, length 40 mm, which are inserted in steel tubes having the following dimensions: wall thickness of 1.5 mm, inner diameter of 10 mm, and lengths varying from 10 to 200 cm. The longer tubes can be conveniently driven into the ground at the site without causing any disturbance; the short tubes (10 cm) are used at complex sites, e.g., brick buildings. The dosimeters are left in position for 4 to 12 months. The phosphor exhibits significant fading: 7 to 15% during a 4-month summer period and 1 to 7% during a similar winter period. Corrections for fading are made by including samples in each probe that have been given a large dose: 50 mGy. When the probes are retrieved, the fading is estimated by comparing the response of the control samples with that of samples freshly irradiated with the same dose.

A calibration experiment similar to that described for the Oxford system was carried out. The irradiation unit was a cylindrical container with a diameter of 80 cm, height of 94 cm, and filled with the uranium and thorium-bearing rock lujavrite from Kvanefjeld, Greenland. The concentrations of radionuclides, determined by gamma-ray spectrometry, were 1550 ppm Th, 552 ppm U, and 4.3% K, and the calculated infinite-matrix dose rate[38] was 140 mGy a^{-1}. The measured dose rate was about 20% less because of radon escape. In the experiment, $CaSO_4$:Dy in steel tubes was compared with Al_2O_3:Si,Na made by Mehta and Sengupta.[110] The response of Al_2O_3:Si,Na is almost identical with that of quartz (Figure 8). In order to cut off beta radiation, the aluminum oxide was contained in quartz tubes with an inner diameter of 2.5 mm and wall thickness of 2.45 mm. The absorbed dose received by the two phosphors in the irradiation unit was estimated from calibration with a ^{60}Co gamma source. The result of the experiment was that $CaSO_4$:Dy shielded by 1.5-mm steel overestimated the dose to Al_2O_3:Si,Na in quartz tubes by 5%. However, when attenuation by the walls of the quartz tube is taken into account,[216] the overestimate is reduced to 3%, i.e., the response of the $CaSO_4$:Dy dosimeter is virtually the same as that of quartz. The approximation to the response of feldspars is even better.

5. Portable Scintillation Counters

As it is not always practical to place TL dosimeters on a site, portable scintillation counters equipped with sodium iodide detectors are now frequently used for measuring environmental radiation. Compared with TL there is the disadvantage that a larger auger hole is required (diameter of 9 to 10 cm) and that no averaging of dose rate takes place. However, there is less disturbance of the work on a site, because the measurements can be completed in a single day and there is the same advantage of automatic allowance for soil inhomogeneity although some differences in response must be kept in mind.[80]

The scintillation counter can be used as a spectrometer to measure separately the contribution from the three gamma sources and cosmic radiation by utilizing the 2.65-MeV gamma emission from ^{208}Tl for the thorium series, the 1.76-MeV emission from ^{214}Bi for the uranium series, and the 1.46-MeV emission from ^{40}K for potassium.[63,80] Alternatively, the total dose rate can be estimated by integrating all gamma radiation above a certain energy by setting an energy threshold on the instrument. Because the relative abundance of Th, U, and K may vary considerably with soil type, a threshold setting must be selected that minimizes the dependence of response on the relative concentration of the three components. Løvborg and Kirkegaard[111] calculated an optimum threshold position of 0.37 MeV for their 3 × 3 in. crystal used for determination of surface dose rate in air above geological formations. Murray et al.,[108] using a 4π geometry, determined experimentally an optimum setting of 0.45 MeV for a 1.75 in. diameter × 2 in. high crystal. As the three components have almost identical variations with energy in the interval of 0.37 to 0.45 MeV, the choice of threshold in this interval is not critical. However, the position chosen must be kept fixed and the energy calibration of the counter must be checked constantly because there is a rapid change of count rate with threshold setting in this interval.

Portable scintillation counters for environmental measurements and their calibration have been described by Murray et al.,[108] Bell et al.,[112] and Liritzis and Galloway.[113]

6. Relation of Soil Dose to Sample Dose

The methods discussed above will give the gamma dose rate at a point in the soil surrounding the sample to be dated. This dose rate may be different from the average dose rate in the sample for two reasons: (1) the gamma radiation from the soil will be attenuated by the sample and (2) the sample will absorb a certain fraction of the radiation emitted by its own radionuclides, called the gamma self-dose.

If the sample and soil have identical concentrations of radionuclides (and the same absorption characteristics), the two effects will cancel and the soil dose will equal the sample dose, i.e., the absorbed *fraction* of gamma radiation emitted by the sample equals the *relative* attenuation of the radiation from the soil. If, on the other hand, the concentrations of radionuclides in the sample and soil differ markedly, a correction may be necessary.

Such corrections require a knowledge of absorbed fractions of gamma-radiation. Values for a unit-density soft-tissue medium were obtained by Brownell et al.[114] and Ellett and Humes[115] by means of Monte Carlo calculations. The values can be utilized for other densities and other low-Z media by applying conversion factors depending on mass and electron densities. Values for ceramics were obtained by McKerrell and Mejdahl.[83] Mejdahl[82] has calculated values for ceramics and stones utilizing the same conversion procedure and the distributions of photons emitted in the uranium and thorium series listed in NCRP (1977).[116]

7. Variation of Gamma-Ray Intensity with Time

The gamma dose rate in soil exhibits a seasonal fluctuation caused by the varying content of water in the soil. A high water content causes an appreciable attenuation of the radiation. At the same time the water prohibits the escape to the atmosphere of the two gases radon and thoron because of the reduced diffusion coefficient.[117] Conversely, in dry periods the gases escape but now the attenuating effect of the water is not present. The two effects are of the same order of magnitude and tend to compensate, with the result that the uncertainty caused by seasonal fluctuations is tolerable. Using TL probes, Mejdahl[109] measured the

Table 6
**SOURCES OF ERROR AND
ESTIMATED UNCERTAINTIES
(1σ) IN THE RISØ TLD SYSTEM
FOR MEASUREMENT OF
ENVIRONMENTAL DOSE RATE
IN POTTERY BURIED IN SOIL**

Source of error	Uncertainty (%)
Measurement of TL	2
Calibration of ^{60}Co source	2
Fading correction	2
Comparison with response of quartz	3
Conversion from soil dose rate to dose rate in pottery	3
Seasonal fluctuations	2
Total uncertainty	6

radiation in the soil (depth 50 cm) at a particular location over a period of 2 years, changing the probe each month. The monthly variation did not exceed 3.5%. The period included extremely dry and extremely wet months, as well as months in which the ground was covered by snow. The soil had a typical concentration of radionuclides, i.e., uranium (in equilibrium) contributed about 25% of the gamma dose. Liritzis and Galloway,[118] using scintillation counter measurements, attempted to correlate variations in gamma dose rate with meteorological factors. They found that the variation of results did not exceed 5.2%. There might be a larger variation at greater depths because the migration of radon and thoron will be different. This problem has been considered by Fleming.[68]

Leaching or precipitation of radionuclides, in particular uranium compounds, may occur in particular soils, e.g., podsols and soil that has been flooded by water containing carbonic acid, for instance from a well. In such cases, the present-day gamma dose rate can differ considerably from the average dose rate during the period of burial. An example is discussed by Mejdahl.[107] Long-term variations can result from the effect of initial disequilibrium in the uranium series as discussed for beta radiation. The effect is even more serious for gamma-radiation because 95% of the gamma-radiation is post-radium. Assuming a uranium contribution of 25%, the maximum error in the total gamma dose rate would be 12%.

8. Uncertainty in the Measurement of Gamma Dose Rate

The total uncertainty in the measurement of gamma dose rate by means of TL dosimeters has been estimated by Mejdahl.[88] The results obtained are listed in Table 6.

The effective uncertainty contribution will be less than 6% because only part of the radiation is contributed by the environmental radiation. A representative value, corresponding to a contribution of 40%, is 2.5%. The uncertainty in scintillation counter measurements will be slightly higher because no averaging of seasonal fluctuations takes place, but it is estimated to be less than 7%.

IV. ARCHAEOLOGICAL AND GEOLOGICAL DOSE DETERMINATION

A. Heated Materials

1. Methods of Additive Doses

The simplest method of obtaining the past radiation dose would be to compare the natural

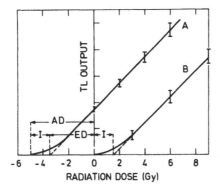

FIGURE 9. Additive method of evaluating the archaeological dose (AD). (A) The primary build-up curve obtained by adding laboratory doses to the natural dose. (B) The secondary build-up curve obtained from samples that have been heated to 500°C prior to the laboratory irradiation. AD is the sum of the equivalent dose ED and the intercept I.

TL signal with that induced in the same sample by a subsequent irradiation. Unfortunately, this method is rarely applicable since changes in sensitivity occur after heating to 500°C to obtain the glow curve. This difficulty rapidly led to the development of the "additive method" which is used for both fine-grain[66] and inclusion[67] dating. This involves the use of at least three equal samples, one of which is used to obtain the natural TL and the other two the two higher levels of TL as a result of two different irradiations prior to heating. This is shown in Figure 9, curve A. Linear extrapolation of the three points onto the dose axis gives the equivalent dose (ED).[11] Provided that the response is completely linear, the ED will be the archaeological dose (AD). However, it is widely known that the initial dose response of most fine-grain and quartz samples is nonlinear, with the sensitivity increasing to a constant value when a dose of the order of 1 Gy is attained. This phenomenon is known as supralinearity and though several mechanisms have been put forward to explain it, none has been proved to be the sole cause.[68,119-121] A comprehensive discussion of supralinearity in LiF Mg, Ti is given in Volume II, Chapter 2.

To take account of supralinearity a second curve B is constructed using samples which have had the TL due to the archaeological dose removed by heating to 500°C. This is called the second-glow growth curve. It is then assumed that extrapolation onto the dose axis will give an intercept, I, which allows for the nonlinear growth and which is considered to parallel the initial build-up of the TL in the archaeological situation. Hence the archaeological dose is given by the equation:

$$AD = ED + I \qquad (13)$$

Supralinearity has been found for quartz and feldspar inclusions as well as for fine-grain samples. However, the simple correction described above is often inapplicable as it appears that the value of I obtained depends upon the value of AD. This can be seen in Figure 10 where the second-glow growth curve has been generated by using samples which have received an additional beta dose prior to the removal of the TL due to the archaeological dose. Fleming has suggested using a correction factor assuming a linear connection between the change of supralinearity and the radiation dose.[68] Uncertainty in the value of I can be an important source of error in the value of AD.

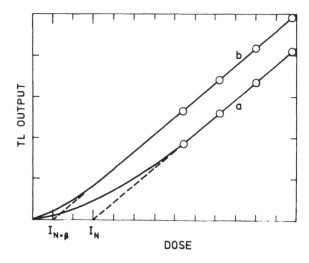

FIGURE 10. Dose dependence of the supralinearity intercept. Curve a is a second-glow response curve generated after drainage of the natural TL and yielding an intercept of I_N rad. Curve b is a second-glow response curve generated after drainage of (natural + laboratory-induced) TL, and yielding an intercept of $I_{N + \beta}$ rad. The slopes of both curves in their linear portions are drawn as equal, representing an idealized situation. (From Fleming, S. J., *Archaeometry*, 17, 122, 1975. With permission.)

Different problems due to nonlinearity are encountered at higher doses as saturation of the TL traps is approached. For the dose levels experienced by most pottery samples this is usually not a problem. However, saturation has been found when the TL technique has been extended to dating baked clay from prepottery cultures and to geological materials. In particular, saturation was found for quartz inclusions extracted from baked sediment[122] and from baked clay-sand ovenstones[123] from Lake Mungo, Australia. The latter applied an exponential fit to the growth curve of the form (1- exp(- x/B)) where x is the dose and B is a constant determined experimentally. The former used a larger number of smaller additional doses so that the TL was kept to the linear part of the curve, i.e. to doses less than 100 Gy.

The same equation was applied to quartz extracted from a granitic inclusion taken from a lava flow in the Massif Central, France;[59] an age of $(34.6 \pm 3) \times 10^3$ years was obtained for the Laschamp flow. For quartz pebbles found in the sediment beneath a nearby flow at Olby, Valladas and Gillot[124] also found nonlinearity for doses greater than 100 Gy; however, for three of the four samples a better fit was found using the expression (1-exp(- X/B))$^\alpha$, where $\alpha > 1$. In some cases this expression could not be used and a more complex normalization procedure had to be adopted. Feldspar has much higher saturation values than quartz. A semiempirical expression (1-exp(- X/B))$^\alpha$ was proposed[125] and used for obtaining geological doses in the range of 1000 Gy for plagioclase feldspars of volcanic origin.

2. Transfer Methods

Conventional additive dose methods have run into various difficulties due to such phenomena as supralinearity, spurious luminescence at high temperatures (where TL signals tend to be weaker), and anomalous fading. As a result of these problems, a new method of dose determination was tried out in the late 1970s. The method uses phototransferred TL (PTTL) to look at the trapped electron population of an energetically deep trap (donor trap) by optically transferring a fraction of them to a shallower trap (acceptor trap) which empties at a lower temperature in the glow curve. It is the signal from this lower-temperature trap which is measured and used to plot growth curves for the dose determination. Ideally the illumination used for the measurement should empty out less than 1% of the trapped electron population of the donor traps and can be given in a time that is short enough for there to be no thermal decay of the electrons in the acceptor traps.

Two techniques have been considered in detail. The first is "additive", involving a single sample; this method would obviously be advantageous for small archaeological samples. Here the PTTL of the natural radiation exposure is obtained and then a laboratory dose is added. The resulting lower peaks are annealed before the PTTL is measured again. This procedure is repeated several times. The second technique, known as the "total bleach" method, involves prolonged UV exposure prior to refilling of the donor traps by laboratory irradiation. The PTTL of the natural radiation dose is thus compared with that due to laboratory irradiation.

The first application of PTTL was to materials that could not be dated by conventional methods because the TL glow peaks in the 300 to 500°C region showed anomalous fading.[33,60] Using 320-nm UV illumination produced by light from a 300-W xenon lamp passed through a monochromator, Bailiff[126] was able to demonstrate that a PTTL signal in zircon could be obtained from traps which did not fade. These donor traps were shown by thermal annealing experiments to be emptied by heating to temperatures above 500°C and are thus not connected with the TL peaks that are known to fade. Using the 320-nm illumination, he showed that the dose response registered by the acceptor trap, which has a peak at 135°C, is linear with dose up to 200 Gy; this dose region includes that of archaeological interest. The study of a deeper PTTL peak in zircon (at 250°C) using shorter wavelength illuminations indicated that more than one donor trap was involved in the PTTL process. Another complicating factor is the presence of a "residual" signal when an annealed but unirradiated sample is exposed to the standard illumination.[127] Similar PTTL studies on fluorapatite were not as promising as those on zircon.

Another mineral which has been extensively studied is Norwegian alpha-quartz, which was thought to have radiation properties similar to those of archaeological quartz extracted from pottery. The incentive for PTTL studies in this mineral is the avoidance of the so-called "malign" behavior of the 325°C peak[67] in which the second-glow response shows an increase in sensitivity and a peak shift relative to the first-glow response. Bowman[128] studied the PTTL of the 110°C peak which is due to the transfer of charge carriers from the 325 and 375°C peaks. The behavior of the PTTL has been shown to be extremely complex, particularly with regard to using the additive method.[128] The behavior in Norwegian alpha-quartz was found to be quite different to archaeological quartz and no routine procedure for using PTTL could be established. However, using the additive PTTL method, Sasidharan et al.[129] did obtain reasonable agreement with the values obtained by the conventional quartz-inclusion method. They used an unfiltered germicidal lamp for the illumination.

Another category of materials that would benefit from a PTTL technique is that which alters its structure on heating during the course of a glow curve, e.g., aragonite changes to calcite, or that which undergoes a large change in sensitivity after first-glow measurements because it has not previously been heated, e.g.,sediments. For these samples, apparatus was developed to carry out PTTL measurements at liquid nitrogen temperatures[130] and good agreement was obtained for an ocean sediment sample using both PTTL dating methods. However, no consistent dose dependence could be found either for aragonite shells or corals.[131]

At the moment, it seems unlikely that any PTTL method will become routine for dating. The wavelength response is complex and not well understood and there seems to be considerable variation in the response of otherwise typical archaeological samples.

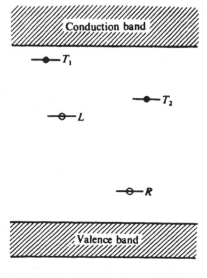

FIGURE 11.   Model for the predose effect
in quartz. (From Aitken, M. J., *PACT*, 3,
319, 1979. With permission.)

3. Predose

A totally different approach to TL dating has been developed for quartz-bearing pottery
less than about 1000 years old. Instead of observing the TL from a deep trap, whether by
direct measurement of a glow curve or indirectly by a phototransfer technique, the predose
method uses the sensitization of the 110°C peak of quartz brought about by heating the
irradiated quartz to 500°C during the course of a glow curve. The possibility of developing
a dating technique based on a dose-dependent sensitivity was proposed by Fleming.[21,132]
The sensitivity increases obtained for the 110°C peak are far greater than those encountered
in dosimeters such as LiF (TLD-100); a typical enhancement factor, reported for an Etruscan-
style terracotta, was $\times$ 0.3/10^{-2} Gy.

As a result of studies on Norwegian alpha-quartz,[133,134] a model was proposed involving
two traps and two hole centers (Figure 11). Trap T_1 is the shallow trap and L is a luminescence
center; an electron released from this trap which recombines at the luminescence center gives
rise to the TL of the 110°C peak. R is another hole trap which is much closer to the valence
band and acts as a reservoir of holes. T_2 is a deeper trap which is not emptied by heating
to 500°C, but which is introduced to maintain charge balance and does not actually enter
into the predose mechanism. The nomenclature is that of Aitken,[135] who gave a compact
outline of the model and predictions based on it. Firing the pottery is thought to empty all
the centers so that the initial sensitivity, S_o, of the 110°C peak, as would be measured by
a test dose of about 10 mGy at that time, is low. Exposure to the natural radioactivity from
that time on produces electrons in the conduction band and only those that get trapped at
T_2 centers are stored; electrons trapped at T_1 centers will decay rapidly at ambient temperature.
The holes that are simultaneously produced in the valence band are assumed to be trapped
predominantly at R rather than L centers which, it is proposed, have a lower capture cross
section.

When S_o is measured in the laboratory, the test dose creates a small number of electrons,
a proportion of which get trapped at T_1; a small number of holes get trapped at R, but this
changes the total population of holes at R by an insignificant amount. On heating to 150°C,
the electrons at T_1 are released and a fraction of them will recombine at the luminescent L
centers and give a TL signal. The sample is then heated to 500°C when holes in R are

released and are captured at L centers. Hence, when the sensitivity is remeasured using a similar test dose, an enhanced sensitivity, S_N, is measured which is proportional to the number of holes that had been trapped at R. The sensitivity can then be further enhanced by irradiating with a dose, β, which will put holes in R, and then heating to 500°C to transfer the holes from R to L. The new sensitivity, $S_{N+\beta}$ will be proportional to the number of holes now trapped at R which will be due to the combined effect of the natural radiation dose and the laboratory dose, β. Hence the archaeological dose is given by

$$AD = \frac{S_N - S_0}{S_{N+\beta} - S_N} \times \beta \tag{14}$$

The age is then given by

$$Age = \frac{AD}{D_\beta + D_\gamma + D_c} \tag{15}$$

as in the case of quartz-inclusion dating. Fleming[68] suggests that more reliable predose behavior is obtained if no HF etching procedure is used. However, this is not a universally accepted view (see, e.g., Wright[136]).

Needless to say, a variety of problems are encountered in practice when archaeological samples are studied. These have been described in detail by Aitken.[135] The problems include nonlinearity caused by "radiation quenching" for which allowance can be made by extra measurements and nonlinearity due to saturation of either L or R centers.[137,138]

Further problems may be caused by partial activation of the predose mechanism at ambient temperature.[139]

B. Unheated Materials

1. Differences from Heated Materials

The recent extension of the TL-dating technique to the dating of unfired sediment has led to the development of slightly different methods of dose determination.[140] The main difference between pottery and sediment comes about, of course, from the lack of heating of the latter; instead it is proposed that the trapped electrons due to the earlier geological radiation exposure were bleached by sunlight prior to sedimentation. This lack of heating has two main effects on the TL of sedimentary minerals. First, the optical emptying of the traps is not as efficient as the thermal emptying so that there is a "residual" TL signal left in the sample at the time of deposition. Second, the material undergoes a major sensitivity change on heating in the course of a glow curve and hence the natural TL cannot be compared with the second-glow response. This also makes it impossible to use the second-glow response to assess the linearity for dose levels below the geological dose for any particular sample.

Other differences are concerned with the dosimetry that is applicable to a sediment. The main advantage of working with sediments is that they are more homogeneous than pottery so that radioactivity analyses of small subsamples are indeed representative of that particular deposit. Another difficulty of working with sediments is connected with the high water content, e.g., the saturation water content is around 40% for loess. Due to uncertainty in earlier times, water content is thus a major contribution to the error assessment of the TL data produced.

A third difference is the simple fact that most of the sediments that require dating are those beyond the range of the radiocarbon technique and therefore far older than any pottery samples. This means that nonlinearity of the growth curve may occur due to the onset of saturation. Fortunately, the feldspars, which are mainly responsible for the TL in most

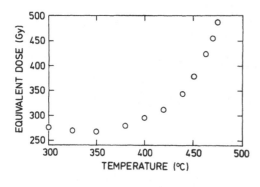 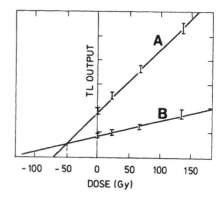

FIGURE 12. (Left)   Equivalent dose ED vs. glow curve temperature for a Norwegian glacifluvial feldspar sample (Risø TL No. 823801, F13). FIGURE 13. (Right)   Growth curves of loess samples at the 275°C glow curve temperature. (A) Natural + additional beta doses. (B) Natural + additional beta doses + 30-min sunlamp bleach. (From Wintle, A. G., *PACT*, 6, 486, 1982. With permission.)

sediments, have a higher saturation level than sedimentary quartz.[141] The greater ages also mean that it is more difficult to find well-dated material for testing the TL method.

2. Partial Bleaching

For young sediments, less than about 25,000 years old, a significant portion of the natural TL signal, I_{nat}, is due to the residual signal, I_o. This is particularly true at higher glow curve temperatures and can be seen if the additive method described in Section IV. A. 1 is applied directly to sediment samples (Figure 12). The natural TL of a sediment sample can thus be divided into two components, $I_{nat} = I_o + I_d$, where I_d is that due to the radiation dose since deposition. Several methods have been used in attempting to separate the two components; these have been reviewed by Wintle and Huntley.[140]

One technique which can be applied, provided that the TL growth is linear with dose, is known as the partial bleach or R-Γ method.[142] In this technique the reduction R, in TL, caused by a fixed bleaching exposure in samples that had received different gamma doses, Γ, in addition to the natural dose, is measured and plotted against Γ. Extrapolation of this R-Γ plot onto the dose axis gives an intercept which is taken to be the geological dose. Another way of plotting, shown in Figure 13, also enables the shape of the residual component I_o to be observed. The R-Γ method is based on the assumption that a fixed fraction of the radiation-induced TL is bleached irrespective of the magnitude of the radiation dose used. This was shown to be the case for some ocean sediments.[71] The method has been used to obtain ages for the polymineral fine-grain component of loess deposited around the end of the last glacial period, some 20,000 years ago.[143] The same method was also applied to the quartz fine grains and although a longer bleaching time was required, the geological doses were in agreement.[141] This method cannot be applied, however, if the N + γ growth curve is nonlinear, either due to apparent saturation of components of the TL signal or to peak shifts caused, perhaps, by second-order kinetics.[141,144] An advantage of the technique is that, provided that the laboratory bleach is short compared with the original sunlight bleach, it could be used for sediments which were not as well exposed to sunlight as loess, e.g., fluviatile deposits.

3. Total Bleaching

A simpler approach has recently been used on some fine grains extracted from stabilized sand dunes in India.[145] They considered that a 1000-min exposure to a 300-W sunlamp would be sufficient to reduce the natural TL level, I_{nat}, to a value negligibly different from the original residual TL level, I_0. They constructed an additive dose curve for the sample and obtained the dose, $D(I_{nat})$, which was the intercept on the dose axis. The geological dose, $D(I_d)$, was given by the equation

$$D(I_d) = (1 - R) D(I_{nat}) \tag{16}$$

where $R = I_0/ I_{nat}$ is the fraction of the natural TL due to the residual TL obtained by the 1,000-minute bleach. With this method reasonable plateaux were obtained for $D(I_d)$ as a function of temperature for 8 samples which subsequently gave ages ranging from 2,000 to 20,000 years. This method avoids the application of irradiations after optical bleaching in the laboratory, but as with the R-Γ method, it can be applied only to samples which are known to have a linear TL response. It also assumes that the UV-rich sunlamp spectrum does not have different bleaching properties from those of the sun.

For fine-grain samples from loess older than about 20,000 years, for which the additive dose method could not be applied because of nonlinearity, a total bleach was applied and the dose was obtained by comparing the natural TL with the TL regenerated by subsequent laboratory irradiation.[146] Further study of the spectral response and the effect of different bleaching times are needed in order to justify this approach.

V. METHODS FOR AGE DETERMINATION

A. Fine-Grain Technique

In the fine-grain technique, developed by Zimmerman,[66] the material used for dating consists of polymineralic grains in the 1 to 8-μm range, a size for which the attenuation of alpha radiation is virtually negligible. In the case of pottery, the fragment is crushed by slowly squeezing it in a vise; the desired grain size is selected by suspending the crushed material in acetone and making use of the dependence of settling time on grain diameter. After separation, the selected grains are resuspended in acetone and deposited on aluminum discs (10 mm in diameter and 0.5-mm thick) in a thin layer of a few microns. About 20 such discs are prepared from each sample. Normally, the TL reproducibility of the discs is within ±5%. The technique has been described in detail by Huxtable[147] and Fleming.[68]

The TL glow curve of fine grains does not exhibit sharp individual peaks, but consists of rather broad bands of light emission,[66] presumably because the sample is polymineralic. Singhvi and Zimmerman[148] identified the luminous minerals in fine grains from a variety of ceramic samples using a scanning electron microscope with cathodoluminescence and X-ray microprobe attachments. They concluded that the minerals responsible for TL were predominantly quartz and feldspar. A peak in the 350 to 400°C region appeared usually to be from quartz. Samples in which feldspar was predominant had a steadily falling glow curve in this region. Apatite and zircon were detected, but the proportion appeared to be too small for the TL contribution to be significant.

Fleming[68] gives a recipe for removing clay grains from a fine-grain sample: boiling in a mixture of concentrated nitric, sulfuric, and perchloric acids (in proportions 1:1:2) for about an hour. The treatment can increase the TL signal by a factor of 5 and helped to reduce spurious TL and anomalous fading found in some samples.

B. Quartz-Inclusion Technique

In the quartz-inclusion technique, developed by Fleming,[67,68,149] quartz grains in the size range of 90 to 150 μm are selected for the TL measurement. Beta radiation from the clay matrix suffers only a small attenuation, less than 10%, in grains of this size, while alpha adiation is severely attenuated. The dose from alpha radiation is eliminated by etching away

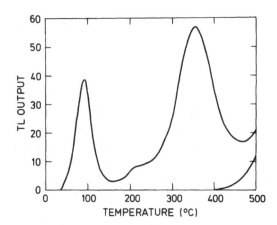

FIGURE 14. Typical glow curve for quartz extracted from pottery. Heating rate is 6°C sec^{-1}. Risø TL No. 810202.

the outer layer of the grains with concentrated hydrofluoric acid. Treatment for 40 to 60 min removes a layer of 6 to 8 μm. It was assumed originally that the acid attack of a grain would be uniform, but in an investigation of quartz from the baked clay of aboriginal fireplaces in Australia, Bell and Zimmerman[74] found that the acid attack proceeded in a nonuniform way for a large proportion of the grains (about 75%) leaving deeply etched pits and giving the grains a "frosty" appearance while the remaining part appeared "shiny". Bell and Mejdahl[150] showed that grains extracted from ceramics fired at a high temperature (e.g., bricks) are less susceptible to acid attack and suggested that the channels of attack would become blocked by impurity atoms diffusing into the quartz grains. The dosimetric effect of this difference in susceptibility to acid attack is probably not severe because one can assume that a minimum of 6 to 8 μm will be removed in any case. However, the resulting variation in grain transparency has consequences for the comparison of beta and gamma doses (see Section V.F).

The TL glow curve for quartz in the 0 to 500°C region (Figure 14) has a distinct peak at 110°C and two partly overlapping peaks at 325 and 375°C (for a heating rate of 6°C sec^{-1}). The 110°C peak is the basis of the predose technique discussed in Section IV.A.3. Only the 375°C peak is used in the normal quartz inclusion method because the 325°C peak most often exhibits a complex behavior, being subject to sensitivity changes and predose effects.

Bell et al.[112] have used larger grains of quartz (up to 0.5 mm) with good results. For still larger grains the attenuation of beta radiation becomes severe and because of grain-to-grain variation it is necessary to grind the grains to a smaller size in order to obtain a homogeneous sample; however, this introduces a spurious TL signal that cannot easily be eliminated.

The 375°C peak can be used for measuring doses in the 2- to 200-Gy range. At the upper end, saturation sets in and consideration of trap creation may be necessary.[68] Assuming an annual dose of 2.5 mGy, the dose range stated above corresponds to an age range of 1 to 80 ka. Archaeologically, the quartz-inclusion method reaches well into the Palaeolithic period and it is also used for the dating of sediments.

C. Dating Based on Feldspar Inclusions

The use of feldspar minerals for dating has been described by Guérin and Valladas,[76] Mejdahl and Winther-Nielsen,[30] and Mejdahl.[82] Feldspars occur in abundance together with quartz in ceramic materials, burnt stones, and sediments. Separation of minerals has been accomplished by means of heavy liquids made by mixing the two organic liquids tetrabromomethane (density 2.96 g cm^{-3}) and dipropylenglycol (density 1.03 g cm^{-3}) in different

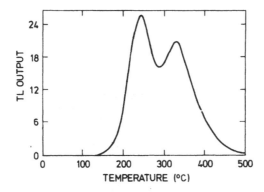

FIGURE 15. Natural glow curve for alkali feldspar
from a Norwegian glacifluvial sediment. Heating rate is
6°C sec⁻¹. Risø TL No. 823802.

proportions. Alkali feldspars (density 2.55 to 2.63 g cm⁻³) can be extracted by using a
liquid with a density of 2.60 g cm⁻³. Plagioclase feldspars having densities in the range of
2.62 to 2.72 g cm⁻³ cannot easily be separated from quartz (density 2.65 g cm⁻³) using
this method. However, because the TL signal from feldspars is 10 to 50 times that from
quartz, it is often possible to use the mixture of quartz and plagioclase feldspars without
separation. A promising magnetostatic method of separation[151] has been tested by Janér and
Jungner.[152]

All sizes of feldspars, from 0.1 mm and upwards, can be used for dating. Grains in the
range of 0.1 to 0.5 mm are treated with 10% HF for 40 min for removal of the outer layer
that has received alpha dosage. The alpha dose contribution in larger grains is negligible.
Grains larger than 0.5 mm are ground down to 0.1 to 0.3 mm in order to obtain a homogeneous
sample. The spurious TL induced by the grinding is removed by treating the ground sample
with dilute HF and HCl (10% solutions) for a few minutes.[82]

Feldspars have a variety of partly overlapping glow peaks in the 50 to 450°C range. A
peak occurring at 650°C in plagioclase extracted from lava was utilized by Guérin and
Valladas[76] in an important method for dating lava flows. A typical natural glow curve for
an alkali feldspar from a Scandinavian sediment (natural dose about 1000 Gy, heating rate
6°C sec⁻¹) is shown in Figure 15. There are two prominent peaks at 230 and 320°C and an
indication of a peak at 415°C. By using preheating, other peaks in the 300 to 450°C range
can be isolated. The peak used for archaeological dating occurs most frequently at 375°C
(at a heating rate of 16°C sec⁻¹). Following laboratory irradiation, there is a prominent peak
at 140°C.

Because of the anomalous fading demonstrated by Wintle[33] for feldspars extracted from
lava (see Section II.A), the use of feldspars for dating has been regarded with some suspicion.
Extensive fading tests of feldspars from archaeological samples have been carried out by
Mejdahl.[82] A few samples showed fading of up to 12% in a 4-week period, but the majority
of samples showed no significant fading. The experience from the dating of a large number
of Scandinavian archaeological samples[30,82] is that feldspar dates are generally in good
agreement with quartz-inclusion dates and dates obtained by other methods. Also in that
reasonable ages can be obtained for feldspars from sediments,[153,154] fading appears to be
ruled out as a serious source of error.

Feldspars have several advantages over quartz. The brightness of the TL signal, 10 to 50
times that from quartz, has already been mentioned. Although the peak used for dating may
be composed of several overlapping peaks, the age plateau is usually very long, frequently
100°C or more in archaeological samples and up to 200°C in feldspars from sediments.[154]

Perhaps the most important advantage is the ability to use large grains of alkali feldspars for archaeological dating. For grains larger than 1 mm, containing a minimum of 10% potassium, more than 50% of the dose is contributed by beta radiation from inherent potassium. As a result, the errors associated with the measurement of environmental radiation and external beta radiation are reduced considerably. This is particularly important in dating burnt stones, which generally have a nonuniform distribution of radionuclides and come from a complex environment. Granitic stones often contain feldspars of several millimeters in which inherent potassium may contribute up to 90% of the dose. The dating of burnt stones using alkali feldspars has been described by Mejdahl.[82]

Feldspars have a measurable dose range of about 0.6 to 2000 Gy. The corresponding age range, assuming an annual dose of 3 mGy, is 0.2 to 700 ka. The long range makes feldspar dating a very important method for geological age determination because it covers a range for which there are few other absolute methods.

D. Subtraction Methods

In the techniques discussed above it is necessary to measure the environmental radiation at the site. As a result, the techniques are limited to fresh excavations; objects already in museum collections cannot be dated. For this reason, and because of the complexity of many environments as well as the uncertainty arising from the possibility of long-term variations in environmental dose rate, there has been a great interest in the development of subtraction methods that would obviate the need to measure the environmental radiation.

A subtraction method based on a combination of the fine-grain and quartz-inclusion techniques was developed by Fleming and Stoneham[155] and further discussed by Fleming.[68] The age equation for the two techniques can be expressed as follows:

$$FG = (env. + \beta + a\alpha)A \tag{17}$$

$$I = (env. + f\beta)A \tag{18}$$

where FG and I are accumulated doses in fine grains and quartz inclusions, respectively; env. is the environmental dose rate, α and β are infinite-matrix dose rates from α- and β-radiation, a is the alpha efficiency factor, f the factor for attenuation of beta radiation, and A is the age. Subtraction now yields

$$FG - I = ((1 - f)\beta + a\alpha)A \tag{19}$$

because the environmental dose rate is the same for both grain sizes. Since $(1-f)$ is very small, about 0.07 for 100-μm grains, the age determination now depends almost exclusively on the α dose rate. Because the α dose rate in pottery is particularly sensitive to fluctuations in radon retention and variations in water content, the accuracy of the method is limited to about 12%. Furthermore, this accuracy is likely to be achieved only if FG-I is sufficiently large, $> 0.3FG$. The method has been tested on a variety of shards of known age and in most cases the difference between the subtraction age and the known age was within 12%.[155] Because of its complexity and relatively poor accuracy, the method has not found routine application but it has been of importance in particular cases where the environment was complex or unknown. As an example. the subtraction dating of pottery produced in the earliest Jomon period in Japan (called Ryusenmon) may be mentioned. A radiocarbon date of 13,060 BP (half-life 5730 years) associated with the pottery indicated that the Ryusenmon is the oldest pottery discovered so far in the world. The date was much debated because archaeological estimates varied widely. Fleming and Stoneham[155] obtained a subtraction age of 13,970 ± 1850 years BP and was thus able to confirm the high antiquity of the pottery.

A feldspar-quartz subtraction method, based on large grains of equal size of alkali feldspar and quartz, was proposed by McKerrell and Mejdahl.[83] The age equations for the two minerals are

$$F = (\text{env.} + f_1\beta_s + f_2\beta_F)A \tag{20}$$

$$Q = (\text{env.} + f_1\beta_s)A \tag{21}$$

where F and Q are accumulated doses in feldspar and quartz, β_s and β_F are infinite-matrix beta dose rates in the sample and from potassium in the feldspar grains, f_1 is the beta attenuation factor and f_2 is the absorbed fraction of the infinite-matrix beta dose in the feldspar. Subtraction yields

$$F - Q = f_2\beta_F A \tag{22}$$

The age determination would thus be based solely on the beta dose rate from the potassium in alkali feldspars and all uncertainties associated with the measurement of environmental radiation and sample beta radiation would be eliminated. The difference $F-Q$ can easily be 0.9F for grains larger than 1 mm. However, because of the difficulties in measuring large quartz grains, the method has not yet been realized. It is more likely to be realized with plagioclase feldspars rather than quartz as the second component. The feasibility of using large feldspar grains for dating has been demonstrated by Mejdahl.[82] In any case the feldspar subtraction method can be applied only to either coarse-ware pottery containing large grains or to burnt stones.

A subtraction technique in which the environmental contribution is eliminated by combining samples having different concentrations of radionuclides has been proposed by Plachy and Sutton.[156] The method was used for dating hearths containing burnt stones of granite and quartzite/sandstone. The latter is typically low in radioactivity and thus records essentially the environmental dose accumulated during the burial period. Granites, on the other hand, usually have large concentrations of radionuclides. Granites can then be dated by first subtracting the environmental dose obtained from the quartzite/sandstone from their archaeological dose and then dividing by the difference in internal dose rate. This procedure automatically corrects for any variation in environmental dose rate over the burial period. However, for highest accuracy it is necessary to take into account the slight variation of the environmental dose from one rock to another depending on the rocks and soil in the immediate vicinity.

E. Other Special Methods

1. Zircon Dating

Zircon is commonly found as an accessory mineral in archaeological ceramics. The zircon grains can be extracted by heavy liquid separation of the mineral grains that remain after dissolution of the clay by the HF treatment. Individual zircon grains can then be hand picked for TL measurements. The use of individual zircon grains for TL dating was first discussed by Zimmerman[55] who had become aware of their presence by observing the clusters of tracks on induced fission track maps of ceramics. The high density of tracks formed in the mica detector is due to the high U content of zircon grains and may be up to 1000 times that of the surrounding clay.

This high U content is potentially advantageous for several reasons. First, within the grain the alpha dose contribution is dominant. This means that there is no need to worry about either the radioactive content of the soil or the past water content of the soil or shard. Second, the combined beta and gamma contributions are usually less than 3%; thus, the problems associated with these radiations, such as supralinearity, are not found in zircon grains.

TL measurements are made on single zircon grains since their individual TL sensitivities and U contents vary. Grains greater than 60 μm in diameter are selected and their natural TL is compared with that induced by two ^{241}Am alpha sources placed on either side of the grain. A correction factor is needed for each grain size since the grains are large compared with the range of alpha particles in zircon.[60] The U and Th contents are obtained by induced fission track techniques and the alpha dose rate is calculated as discussed in Section III.A.2.

Successful results were obtained on only 25% of the grains extracted from pottery of a known age; the remaining samples gave TL ages which were too young and this seemed to be connected with "zoning" of the U content and the luminescence sensitivity. Also, some of the grains showed anomalous fading. Hence this technique has not been used very much for dating, but has been applied successfully to a problem of art authenticity. A bronze statuette of a horse, said to have been cast in the 5th century B.C., had been thought for many years to have been a modern forgery. The remains of the quartz casting core were not thought to be suitable for TL measurements since the horse had been extensively X-rayed. However, TL measurements on three zircon grains indicated an age of the order of 2000 years, thus supporting its antiquity.[157]

2. Decay Methods

In the method called DATE (Différence des Atténuations Temporelles des Emission) proposed by Langouet et al.[158,159] the need to measure dose rate is completely eliminated. The method can be used for minerals having two peaks: a peak P_1 whose lifetime is too short for linear accumulation of trapped charges and a high temperature peak P_2 having linear accumulation of trapped charges. The age is determined by comparing the charges accumulated in the corresponding traps.

Assuming that all traps were empty at the time t = 0 (the time of firing for an archaeological sample), the number n_r of traps filled at time t in the absence of fading is

$$n_r = KDt \tag{23}$$

where K is a constant and D is the average dose rate. The number n_{rv} of traps filled at time t if fading takes place is given by

$$n_{rv} = \frac{KD}{p} (1 - e^{-pt}) \tag{24}$$

where p is the probability of escape per unit time. A factor g(t) is defined by

$$g(t) = \frac{n_{rv}}{n_r} = \frac{1}{pt} (1 - e^{-pt}) \tag{25}$$

where n_r is the number of traps of type P_1 that would have been filled in the case of linear accumulation. The ratio n_{rv}/n_r is equal to AD'/AD where AD' is the apparent dose as determined by P_1 and AD is the dose determined by P_2.

If the decay process can be described by first-order kinetics, the probability p is equal to $1/\tau$ where τ is the lifetime of P_1 given by

$$\tau = s^{-1}\exp(-E/kT) \tag{26}$$

where s is the frequency factor, E is the trap depth, k is Boltzmann's constant, and T is the temperature (K). The expression for g(t) becomes

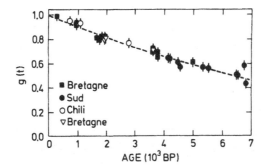

FIGURE 16. Experimental and calculated (- - -) values of the factor g(t) as a function of age t. The calculation was based on a peak lifetime of 4300 years and first-order kinetics. For definition of g(t), see text. (From Langouet, L., Roman, A., Deza, A., Brito, O., Concha, G., and Asenjo de Roman, C., *Rev. Archeometrie*, 3, 57, 1979. With permission.)

$$g(t) = \frac{\tau}{t} (1 - e^{-t/\tau}) \tag{27}$$

The limiting values for g(t) are

$$
\begin{aligned}
&t \text{ very small, } g(t) \sim 1 \\
&t \text{ large } (t \gg 3\tau), \, g(t) \sim \tau/t \tag{28}
\end{aligned}
$$

The latter case corresponds to a state of equilibrium between trapping and untrapping of charges. If a constant burial temperature of the samples can be assumed, g(t) will be a universal function of t for a given pair of traps and once $g_s(t)$ for a particular sample has been measured, the age A can be found by reference to the function (Figure 16).

Langouet et al.[159] tested the method for two peaks in quartz by determining g(t) empirically by means of archaeological samples of known age. Series of samples were obtained from regions having quite different climates; however, almost all measurements could be fitted by a theoretical curve calculated for $\tau = 4500$ years (Figure 16).

In practical applications it may be necessary to build up a master curve, as was done by Langouet et al., because the transitions cannot always be assumed to follow simple first-order kinetics. This will limit the accuracy to perhaps 10%, but the method could still be very valuable in cases where the dose rate has not been constant or is unknown.

A decay method very similar to DATE was proposed by Charalambous and Michael[160] and Charalambous et al.[161] As with DATE the presence of one or more types of traps too shallow to fill at a linear rate at ambient temperatures is required. Assuming that the sample is being irradiated at a constant rate R and a constant temperature T, then the growth rate dI/dt of TL intensity I at the glow peak corresponding to a shallow trap is given by

$$\frac{dI}{dt} = RL - Gexp(-E/kT)I \tag{29}$$

where L is the increase of TL per unit dose (in the absence of decay), G is a constant (for first-order kinetics), E is the trap depth, and k is Boltzmann's constant. If the sample is so old that the peak has reached equilibrium then dI/dt = O and the dose rate R can be determined from Equation 29 when the burial temperature T is known. If the peak has not reached equilibrium then the general solution to Equation 29 can be expressed as

$$I = F(D, Ct) \tag{30}$$

with

$$C = \exp(-E/kT) \tag{31}$$

D is the total dose as measured by means of a stable high temperature peak and t is the total irradiation time. The dating of the sample proceeds as follows. After determination of the total archaeological dose D, the sample is given a laboratory dose equal to the archaeological dose during a time t_1 and at a temperature T_1 such that the intensity of the nonstable peak equals its intensity in the natural glow curve. In that case one has from Equation 30:

$$F(D_1 C_1 t_1) = F(D_1 C_a t_a) \tag{32}$$

where the subscripts 1 and a refer to laboratory and field conditions, respectively. Equation 32 yields

$$C_1 t_1 = C_a t_a \tag{33}$$

or using Equation 31:

$$t_1 \exp(-E/kT_1) = t_a \exp(-E/kT_a) \tag{34}$$

From Equation 34 the age t_a can be calculated when the ambient temperature T_a is known. T_a can be determined if two nonstable trap types are present in the same mineral or in two different minerals from the same sample.

The method has the great advantage that no measurement of the dose rate is required. Two assumptions are made, namely that the TL sensitivity does not change as a result of the first heating of the sample and that the distribution of charges in traps of different depths is independent of dose rate and irradiation temperatures. Only very preliminary tests of the method have been reported so far.

3. Gamma-TL Method

Schvoerer et al.[162] proposed a new version of the fine-grain method, the Gamma-TL method, in which the annual dose from alpha, beta, and gamma radiation is determined solely from measurements of the environmental gamma radiation and the a-value, while the accumulated dose is measured in the usual way. The method was developed for the specific cases (kiln walls, walls or constructions of bricks, heaps of pottery) where the sample to be dated and its surroundings have the same contents of U, Th, and K. A further assumption is that the concentration of radionuclides varies within the ranges of 0 to 5 ppm U, 10 to 20 ppm Th and 1 to 3% K.

A function

$$\Gamma = \frac{I}{I_\gamma} = \frac{aI_\alpha + I_\beta + I_\gamma}{I_\gamma} \tag{35}$$

is defined where a is the alpha efficiency factor and I_α, I_β, I_γ are annual doses from α, β, and γ radiation, respectively. When concentration values U, Th, K and appropriate dose-rate conversion factors are inserted, Γ can be expressed as

$$\Gamma = \frac{(267a + 24)U + (74a + 6)Th + 113K}{12U + 4Th + 27K} \tag{36}$$

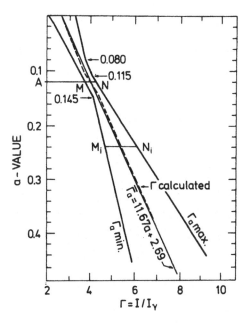

FIGURE 17. Variation of $\Gamma = I/I_\gamma$ as a function of the a-value. I = total radiation intensity; I_γ = intensity of gamma radiation from the soil. (From Schvoerer, M., Bechtel, F., Deshouillers, J.-M., Dautant, A., and Gallois, B., *PACT*, 6, 86, 1982. With permission.)

Assuming equilibrium in the decay chains (in particular no escape of radon) and no attenuation by water, minimum and maximum values of Γ as a function of a can be calculated by varying the concentration values within the ranges stated above. The resulting variation of Γ is shown in Figure 17. The mean value $\overline{\Gamma}_a$ fits a straight line with the equation

$$\overline{\Gamma}_a = 1.67a + 2.69 \tag{37}$$

The method suggested for determining the total dose rate of a sample when the gamma dose rate has been measured is then to measure the a-value, calculate $\overline{\Gamma}_a$ from Equation 37 and insert it in Equation 35. For a-values in the range of 0.08 to 0.145 the error resulting from the use of $\overline{\Gamma}_a$ rather than the actual value of Γ does not exceed 5%.

The method can also be used when the two assumptions of equilibrium and lack of water attenuation are unfulfilled. In fact, a radon loss of 50% would result in an error of only 4% in the total dose rate for typical concentration values when the gamma dose rate is determined by on-site methods. Similarly, it can be verified that the error resulting from the presence of water is small. Results of preliminary dating tests were satisfactory. However, the heavy reliance on the measurement of the gamma dose rate, the most difficult component to measure accurately, would seem to be a drawback of the method.

F. Calibration of Laboratory Sources of Radiation

The natural dose accumulated in archaeological and geological samples can usually be determined with a precision of 5 to 6% evaluated as the standard deviation of results of repeated measurements on aliquots from the same sample. The accuracy that can be achieved in the dose determination rests ultimately on the calibration of the laboratory sources used to construct the build-up curves from which the dose is evaluated.

The primary reference source is usually a ^{60}Co or ^{137}Cs gamma emitter, calibrated by comparison with an international standard. The exposure rate at a fixed distance from the source can be reproduced with an accuracy of about 1.5%. In practice, a ^{90}Sr/^{90}Y beta source is often used for irradiating samples because a sizeable dose rate (3 to 6 Gy min^{-1}) can be achieved without shielding considerations that are too severe. The beta dose rate is determined by comparison with a gamma source by means of TL phosphors such as CaF$_2$ [163] or quartz.[164,165] There are, however, a number of complicating factors affecting the comparison of beta and gamma sources by means of TL phosphors, in particular for the large grains used in inclusion dating; these include (1) backscatter effects, (2) grain-size effects, and (3) grain-transparency effects.

A review of electron backscattering with particular reference to the dependence of the backscatter number coefficient and the fraction of backscattered-to-incident energy on the Z of the absorber and the energy and angle of incidence of the electrons has been given by Kalefezra and Horowitz.[166] They compiled theoretical and experimental data for Al and Si and carried out Monte Carlo calculations for Si and LiF. Murray and Wintle[167] studied the effect of backscattering from perspex, aluminum, nichrome, and lead by measuring the TL from 100-μm grains of CaF$_2$ irradiated on these materials with a ^{90}Sr/^{90}Y beta source. The measurements showed a backscatter of nearly 20% for aluminum and 40% for nichrome relative to the result for perspex.

Using a ^{90}Sr/^{90}Y beta source Wintle and Aitken[168] and Wintle and Murray[169] studied the deposition of beta dose as a function of depth in aluminum absorbers interposed in front of a TL dosimeter consisting of fine grains (1 to 8 μm) of CaF$_2$ deposited onto aluminum discs. The resulting dose depth curve showed an initial increase with depth until a maximum value more than 20% higher than the entrance value was reached at a depth of about 40 mg cm^{-2}, corresponding to about 150 μm in quartz. This initial build-up of dose is explained by an increase in the obliquity of the electron paths due to multiple scattering of the electrons as they penetrate the absorbing medium. Beyond the depth corresponding to the maximum, attenuation effects predominate and lead to a steady decrease of dose with depth. Such build-up and attenuation effects can be expected to cause a nonuniform dose distribution in quartz or feldspar grains, the actual function being dependent on the grain size. For example, in a monolayer of 100-μm grains, corresponding to about 26.5 mg cm^{-2}, a larger beta dose will be deposited in the lower regions of the grains than in the upper ones.

Bell and Mejdahl[150] pointed out that a nonuniform distribution of beta dose would entail a dependence of the apparent beta dose rate on the transparency of the grains when the beta dose rate is estimated by comparing the TL from beta-irradiated grains with that from grains that have received a uniform dose of gamma radiation. For instance, the beta dose rate estimated by means of 100-μm quartz grain would be higher for transparent grains than for the more opaque ones of the frosty category. In a calibration experiment comprising quartz grains from about 100 archaeological samples, the beta dose rate estimated by means of these grains varied from 4 to 6 Gy min^{-1}. TL dating of archaeological samples of known age showed that the TL ages would be grossly in error unless the beta dose rate corresponding to the particular sample was used. The high values of the beta dose rate were found especially for bricks and tiles that had been fired at a high temperature. The temperature dependence of the appearance of the grains (after etching with HF) was confirmed by heating quartz grains in clay at different temperatures. The transparency effect is less pronounced for feldspar grains. Calibration experiments comprising 33 samples gave beta dose rates in the range of 4.0 to 4.8 Gy min^{-1}.[30]

The conclusion to be drawn from the complications discussed above, in particular the transparency effect, is that the establishment of a "once-and-for-all" beta source calibration factor using quartz or any other phosphor is impossible. For utmost accuracy it is necessary to carry out a calibration for each particular sample to be dated by the inclusion technique.

Table 7
TL DATING RESULTS FOR CERAMICS AND
BURNT CLAY FROM A MEDIEVAL KILN
EXCAVATED AT LÜBECK, GERMANY[174]

Lab. no.	TL method[a]	Dose rate (mGy a⁻¹)	AD[b] (mGy)	TL age (a)	TL date (AD)
K 179 A	FG	6.11	4400	720	1258
K 180 A	FG	6.29	4130	657	1321
K 180 B	FG	6.16	4250	690	1288
K 179 T	FG	4.60	3850	836	1142
	QI	3.23	2330	721	1257
K 180 T	FG	5.50	4030	733	1245
	QI	3.19	3170	830	1148
Average TL date		AD 1244	(±26 a, ±60 a)		

Note: Only ED was used in the calculation of the last result (830 a) because
of bad reproducibility of the second glow (Wagner, G. A., personal
communication).

ᵃ FG = fine grain; QI = quartz inclusion.
ᵇ Archaeological dose.

In practice, it might be feasible to use a set of calibration values corresponding to different regions of firing temperature. The total uncertainty of the determination of the archaeological dose when the beta source is calibrated for individual samples is estimated to be about 5%.[88]

VI. EXAMPLES OF RECENT DATING APPLICATIONS

A. Introduction

In recent years there has been a considerable expansion of the application of TL dating to archaeological material other than pottery, e.g., burnt stones, flint and slag, and to geological materials such as lava, calcareous deposits, sand dunes, loess, till, fluvioglacial deposits, and ocean sediments.[170,171] Representative examples of dating applications are given below. A list of dating results obtained at the Oxford laboratory up to 1979 has been compiled by Fleming.[68]

An important problem in age determination is the assessment of meaningful error limits for a TL age. Aitken and Alldred[172] and Aitken[11,173] proposed a system based on two error estimates, the precision p and the absolute error a, both at the 68% level of confidence. The error limit a is the overall error calculated by considering all known sources, both systematic and random. The error limit p is calculated from the statistical spread in individual TL ages A_i obtained for a set of N samples from the same context. It is defined as

$$p^2 = \frac{\Sigma(A_i - A)^2}{N(N - 1)} \tag{38}$$

where A is the average TL age. p is an estimate of the standard error of the average TL age for a context, ignoring systematic errors that are common for the samples from the context; a is a more realistic measure of the uncertainty and is the error limit to be used in comparing TL dates with those obtained by other methods. It is usually within 5 to 7% for archaeological samples.

Table 8
TL DATING RESULTS FOR CERAMICS (RISØ TL NO. 811708) FROM A VIKING AGE HOUSE EXCAVATED AT LEJRE, DENMARK

Grain size (mm)	Dose rate (mGy a^{-1})				AD (mGy)	TL age (a)
	Env.[a]	Beta (S)[b]	Beta (I)[c]	Total		
Q 0.3—0.5	0.78	2.98	0	3.76	3360	894
FA 0.1—0.3	0.78	3.26	0.51	4.55	4050	890
FP 0.5—0.8	0.78	2.69	0.28	3.75	3570	952
FA 0.5—0.8	0.78	2.69	1.76	5.23	4600	880
FA 0.8—1.0	0.78	2.42	2.03	5.23	5080	971
FA 1.0—2.0	0.78	1.93	2.77	5.48	5000	912

Average TL age (a) 917; standard deviation 4.0%
Average TL date AD 1060 ($\pm$ 15 a, $\pm$ 60 a)
Note: Q = quartz; FA = alkali feldspar; and FP = plagioclase feldspar.

[a] Environmental gamma + cosmic radiation.
[b] Beta-radiation from the gross sample.
[c] Beta-radiation from inherent potassium.

From Mejdahl, V., *PACT*, 9, 351, 1983. With permission.

B. Archaeological Applications

1. A German Medieval Kiln

Wagner[174] reported the TL dating of a Medieval pottery kiln, Koberg 15, excavated in Lübeck, Germany. Two materials were available for dating: pottery wasters found in the kiln and burnt clay from the bottom of the kiln. Both were assumed to date the same event, namely the abandonment of the kiln. Three samples of shards and two of clay were dated by the fine-grain and quartz-inclusion methods. The water content during burial was assumed to be 0.75 ± 0.25 times the saturation values. The total annual doses (calculated from concentration measurements), the archaeological doses, the ages and the dates obtained are listed in Table 7. The TL date of the kiln, calculated as the average value of the results in Table 7 is

$$AD\ 1244\ (\pm 26\ \text{years},\ \pm 60\ \text{years})$$

where the first term in the bracket is the precision p and the second the overall error a.

2. A Danish Viking Age Settlement

A Viking Age settlement excavated at Mysselhøjgård, Lejre, Denmark was dated using quartz and feldspar grains extracted from pottery and burnt stones.[82] Results were obtained for a variety of grain sizes as shown in Table 8 for a ceramic sample. For alkali feldspar grains larger than 1 mm the dose from inherent potassium becomes predominant. Two ceramic samples and two burnt stones were dated and the resulting TL date for the site was

$$AD\ 1030\ (\pm 25\ \text{years},\ \pm 60\ \text{years})$$

in good agreement with the archaeological estimate: AD 950 to 1050.

3. Indian Site: Sringaverapura

TL dating of pottery from Sringaverapura, India has been reported by Agrawal et al.[175]

Table 9
TL AGES OF FRENCH PALEOLITHIC SITES
OBTAINED AT BORDEAUX[177]

Site	Period	Material	TL age (years BP)	[14]C age
Abri Duruthy	Magdalenian VI (beginning)	Sandstone	11,300	11,350
	Magdalenian IV (middle)	Sandstone	14,500	13,600
Abri Suard	Riss III (bottom)	Pebble	126,000	

Note: Radiocarbon dates (Lyon) are included for comparison.

The site is a mound situated on the bank of River Ganga, 35 km upstream from Allahabad. The excavation yielded archaeological material belonging to seven different cultural periods in a total deposit of nearly 9 m. TL dates obtained by the fine-grain method for seven samples ranged from 2660 to 3015 BP (before present). The dates showed a stratigraphical sequence and were in good agreement with archaeological estimates and with two radiocarbon dates from the same strata. The site could be related to the so-called Ramayana episode and it was concluded on the basis of the TL dates that this episode was unlikely to be earlier than the middle of the 8th century BC.

4. Paleolithic Sites

Considerable progress has been made in the TL dating of heated sandstones and other material from Paleolithic sites. TL ages of three French sites, two Magdalenian, and one Riss III, obtained by Bechtel et al.,[177] are listed in Table 9. Radiocarbon dates for the Magdalenian sites are included for comparison; these are in good agreement with the TL dates. The third TL age may be compared with an amino acid age of 150,000 years obtained for the level Riss III at Arago, France. The uncertainties of the TL ages were estimated to 8.6% for the Magdalenian sites and 12% for the Riss III site.

Burnt sandstones from four Magdalenian sites have been dated by Valladas.[176] The ages obtained were Pincevent, level IV$_4$ (13.8 ± 0.6)ka; Etiolle (15.2 ± 1.0)ka; Marsangy (11.7 ± 0.7)ka; and Verberie (13.3 ± 0.85)ka. The site at Pincevent was included in a study undertaken by Plachy and Sutton [156] with the aim of developing a technique for dating heated granites. A preliminary date from Pincevent was established as 12.0 ± 1.5 ka BP which is in good agreement with four radiocarbon dates ranging from 11,000 to 12,300 BP.

Ichikawa and Nagatomo[178] dated three burnt sandstones from Senpukuji Cave in Japan found in association with early Jomon pottery. Using the quartz-inclusion technique they obtained an age of (11.84 ± 0.74)ka. This age may be compared with a radiocarbon date of (12.70 ± 0.50)ka and the subtraction date of (13.97 ± 1.85)ka obtained by Fleming and Stoneham.[155]

5. Dating of Chert and Flint

Work on the possibility of using heated chert and flint for dating has been reviewed by Wintle.[170] A major difficulty has been the preparation of samples suitable for dating. A technique for the preparation of thin slices was developed, but they often showed poor reproducibility[179] and the dosimetry was complex, involving consideration of optical absorption properties of the slices. More recently, good results have been obtained by crushing the samples and applying either fine grains (1 to 8 μm) deposited onto aluminum discs or grains of a size around 100 μm.[182-186]

Table 10
TL DATING RESULTS FOR BURNT FLINT FROM A BRITISH MESOLITHIC SITE

Sample no.	Archaeological dose (mGy)		Plateau region (°C)	a value	Internal dose rate (μGy a⁻¹)		TL age (ka)	Total error (ka)
	Fine grains	Coarse grains			Alpha	Beta		
220 A	4600 ± 600	4650 ± 500	340—410	0.140	85	91	9.44	0.75
E	4600 ± 500	4800 ± 300	325—450	0.054	76	118	9.12	1.37
F	6100 ± 300	6400 ± 300	350—390	0.051	200	274	7.77	0.70
G	6830 ± 800	6300 ± 500	310—425	0.050	152	218	10.04	1.35
H	3420 ± 200	3750 ± 300	320—410	0.080	51	73	7.87	0.68

From Huxtable, J. and Jacobi, R. M., *Archaeometry*, 24, 164, 1982. With permission.

The two grain categories will have received the same natural dose because of the homogeneity of the material. The spurious signal induced by the crushing can be reduced to an acceptable level by treating the crushed sample with dilute hydrochloric or acetic acid.

Huxtable and Jacobi[185] dated five pieces of burnt flint from a British Mesolithic site, Longmoor Inclosure, East Hampshire, using the crushing technique to obtain fine grains and grains in the range of 90 to 150 μm. The results obtained are given in Table 10. The following features are noted: (1) archaeological doses lower than 5 Gy can be measured with satisfactory accuracy; (2) the ADs obtained for the two grain categories are in good agreement; (3) the plateau extends over a minimum of 70°C and encompasses the main peak at 375°C; and (4) the internal dose rate is very small. The environmental dose rate measured by TLD was 0.30 mGy a⁻¹. A mean TL age was not given because there was a risk that some flints might have been heated at a time subsequent to the initial settlement. However, the TL dates obtained agree well with the radiocarbon age interval, 9000 to 8000 BP, obtained for sites having similar assemblages of microliths.

C. Lava Flows

The most obvious geological material to which TL studies could be applied is volcanic lava. The earliest experimental study of lava samples was reported by Sabels in 1962[187] and he suggested that it might be possible to obtain dates for lava less than a million years old. Subsequent studies on specific minerals extracted from the lava flows gave inconsistent results on known age samples.[188,189] Part of the problem was discovered by Wintle[33] in a study of lavas of known age collected in France and Iceland: the most sensitive TL minerals, the feldspars, showed "anomalous fading". This precluded the direct application of pottery dating methods to volcanic minerals. In the same year Berry[190] reported finding a relationship between age and natural TL signal for several plagioclase extracts taken from Hawaiian basalts. This study led to the development of a relative dating method for plagioclase feldspars from alkali basalts in the Hawaiian Islands.[191,192]

In order that they could obtain absolute dates for lava flows in central France, several authors have used the TL of lava-baked stones and sediments.[59,124,193] These results and other applications to volcanic materials are discussed in a recent review of nonpottery materials.[170] Dates on lava-baked loess and lake sediments have recently been reported in China.[194]

The most interesting development in dating lavas has been the return to the study of plagioclase feldspars extracted from the lava. In 1975, Brou and Valladas[195] described a TL oven which had been specifically designed to observe low light levels emitted by samples heated up to 800°C. It was subsequently used to study TL of feldspars in the 500 to 700°C region.[196] No fading was found in this region and a plateau was obtained for the geological dose for each of a set of samples of known age from the Massif Central[76] The dose rate for 80- to 125-μm plagioclase grains was found to be extremely complex, but the technique is now being applied on an almost routine basis to other volcanic areas in France.[197]

D. Sediments
1. Deep-Sea Sediments

The first reported study of the thermoluminescence properties of deep-sea sediments was that by Bothner and Johnson[198] in which they appeared to be studying a signal from the calcitic remains of foraminifera. No more papers on this subject appeared until Huntley and Johnson,[199] on studying the TL of siliceous ocean sediments, concluded that since the TL signal, and hence the radiation dose, increased with depth down two cores, it might be possible to develop a dating technique. Extension of this preliminary study to single radiolaria soon showed that the TL signal was, in fact, coming from adhering detrital grains. This rapidly led to the realization that another zeroing mechanism, namely exposure to sunlight, must be effective; otherwise all detrital grains would have a very large TL signal, probably in saturation.

Preliminary results were presented in a paper by Wintle and Huntley[71] and details of the method and background experiments were published the following year.[142] The ages obtained for one of the cores agree with the biostratigraphy back to about 130,000 years ago. The ages obtained had been calculated using a time-dependent dose rate to allow for the decreasing contribution from ^{230}Th which decays towards its equilibrium level. The agreement is all the more remarkable since an estimated water content of 57% was used in the calculations for all the samples. The other core had no marker horizons apart from the Brunhes/Matuyama boundary, which was well below the maximum depth for which TL dates were obtained. However, the six TL dates showed a linear increase in age with depth.

TL dating has also been tried on shales taken from a core 60 km off the west coast of India.[200] The TL signal was considered typical of $CaCO_3$. The ages obtained were all about 115 times lower than the paleontological ages and no satisfactory explanation for the difference has been found.

2. Aeolian Sediments

Since optical bleaching has been shown to be an efficient method of bleaching deep-sea sediments, TL workers soon realized that the method was likely to be most applicable to wind-blown sediments. In particular, the wind-blown silt, known as loess, has received considerable attention.[143,201,202] Loess was also the first sediment type to be studied in the Soviet Union. TL dates on loess have been produced at the Institute of Geology in Kiev since 1965. Unfortunately, this work has been carried out without considering the dosimetry at all and using a single X-ray irradiation for calibration. These omissions have recently been discussed in two review papers on TL dating of sediments.[140,203] It was concluded that where agreement was found with other dating methods, it must be considered fortuitous due to the cancellation of many opposing effects. Criticisms of some of the TL ages have also been written by several geologists in the Soviet Union and a review of their papers has recently been published.[204]

Since 1974, various sediment types have been studied in Tallinn and the dosimetric properties of the quartz and feldspar minerals have been of particular interest.[203,205] Application of this knowledge to a 4-m loess sequence at Lihvin yielded 6 ages in correct chronological order which agreed with the geological interpretation of loess deposition over the last 100,000 years.[206] References to TL dates of loess deposits in China, Hungary, and Poland can be found in the review by Wintle and Huntley.[140]

Other aeolian deposits that have been studied are sand dunes.[145] At the moment it seems that aeolian sediments are the most reliable material for dating and can be used for samples up to 150,000 years old.

3. Glacial Sediments

TL dates have also been reported for a number of different types of glacial sediments, such as tills and glaciolacustrine and glaciofluvial deposits. For these samples it is not known whether sunlight is the only zeroing mechanism or whether grinding of the bedrock by the ice could also play a role since it is well known that pressure can reduce the TL signal. Results obtained for a variety of Pleistocene sediments from Spitsbergen[207] and from sections on the East-Europe plain[206] suggest that the finer silts deposited under water are the most suitable glacial sediments. Larger grains are likely to have been transported a shorter distance and therefore will have retained a larger fraction of their earlier TL signal; this would lead to erroneously large TL ages being obtained. Results on a Canadian till have shown a discrepancy between the TL ages obtained for polymineral fine grains and quartz fine grains.[144] Kronborg[154] obtained TL ages for a selection of Danish waterlaid sediments using alkali feldspar grains in the range of 100 to 300 μm. The ages were in good agreement with other age estimates for the sediments, including radiocarbon ages for the younger sediments.

E. Meteorites

Thermoluminescence was first used to study the radiation history of meteorites in Berne in the early 1960s. Since then several different TL projects have been applied to determine various aspects of meteorite chronology. These include

1. The date at which the parent body was formed
2. The date of separation of the meteoroid from its parent body
3. Information about meteoroid orbits
4. Terrestrial age determination

The TL signal from meteorites is extremely complex. The parent body was exposed to cosmic radiation in space ever since it was formed about 5×10^9 years ago as dated by rubidium-strontium techniques. There was little thermal drainage of the trapped electrons while the meteoroid was in space since the ambient temperature is about 120 K. The separation of the meteoroid from its parent body (in space) gave rise to severe shock and potassium-argon, and ^{40}Ar-^{39}Ar dating indicated that this event took place around 5×10^8 years ago. A recent study[208] has confirmed that there is a correlation between the TL sensitivity and the K-Ar age, as had been found earlier by Liener and Geiss,[209] and the results of combined shock and annealing experiments suggested that the TL sensitivity was reduced by the shock when the meteoroid separated. Both these studies involved the TL of peaks above room temperature which are produced by laboratory irradiation at room temperature.

Some information concerning meteorite orbits has recently been obtained by Melcher.[210] While the meteoroid is in orbit around the sun, the TL signal in the region of the 200 to 300°C glow curve temperatures reaches an equilibrium, with the TL production by cosmic rays being balanced by TL loss by solar-induced thermal annealing. The dose rate in space is 30 to 50 mGy a^{-1}. More thermal bleaching will have occurred for those meteoroids with orbits closer to the sun. For such TL studies to be possible, the meteorite must be large enough that one can remove the outer layer which will have been heated by the entry of the meteorite into the atmosphere of the earth. This heating affects only the TL of the fallen meteorite up to 2 cm from the fusion crust.[211] It is not possible to date the fusion crust itself since it has usually been eroded by weathering. The TL equilibrium level was directly related to the orbit of the meteoroid and among the samples studied was one which appeared to have an unusual orbit.[210] This technique can be applied only to recent meteorites where thermal decay of their TL due to the ambient temperature of the earth has not affected the TL signal.

This decay has also been the subject of several recent papers[212-214] and attempts to give terrestrial ages have been made. The low temperature TL (around 200°C) decays with time, and using the ratio of this TL to that at 400°C, they calculated terrestrial ages for meteorites found in Antarctica. The values obtained depended on the TL kinetics assumed for the peaks and further discussion of the application of TL measurements will no doubt be forthcoming.

F. Limitations of Present Techniques

Any TL dating is limited by the response of the samples that is available. Some types of pottery and burnt stones, such as those which contain large crystals of quartz and feldspar, can be dated relatively easily by the routine methods described above. However there are materials, such as shell and bone, which have a large number of different problems associated with the TL measurements, making it unlikely that they will ever be dated by any TL technique.

Dating by TL is rather like solving a murder — the event has taken place without being observed directly. All the evidence is still at the scene of the crime and the laboratory task is to unravel the evidence presented by the different witnesses. In the case of TL, each piece of evidence has its own random and systematic errors associated with the laboratory measurements.[172,173] The major sources of error that limit the technique are those caused by changes that occurred in the past and which are difficult to monitor in the laboratory today. The most important of these are the past water and radioactivity contents. These points are of particular relevance to sediments whose structure is more open than that of a piece of pottery and which are usually older than pottery.

As the technique is extended back in time, two other problems are encountered. First, one must be sure that the peaks used to measure the past radiation dose are stable enough for no electron loss to have occurred. Rough limits can be set on the decay times from kinetic studies on the peaks. Second, the TL process is eventually limited by saturation of the TL sensitivity. Onset of saturation occurs at different doses in different minerals and these properties must be determined individually.

For younger samples the dating accuracy is limited by the assumptions applied in each individual case, e.g., assumptions regarding the distribution of radionuclides in the pottery or regarding the dose dependence of supralinearity. The precision of the date is limited by the large numbers of different measurements (at least ten) that have to be made on each sample.

VII. CONCLUSIONS

When the role of TL in archaeological dating was reviewed 10 years ago,[22] 5 laboratories around the world were engaged in research and application. Today the number has increased to more than 20 and the extension of the technique to nonpottery materials, including burnt flints and stones, calcareous deposits, volcanic lavas, and, above all, geological sediments, has greatly widened the scope of TL dating.

The development of new TL dosimeters and new detectors for radioactivity measurements has increased the ability to measure accurately the dose rate experienced by the samples today. However, the new techniques have also led to a greater awareness of the effect of changes in the radiation environment in the past.

Recognition of the limitations of the TL technique, in particular with regard to the errors inherent in the large number of measurements which must be made to obtain a TL date, has led to a realistic evaluation of the use of TL in archaeology, involving focusing on problems and periods for which the accuracy attainable (5 to 8%) is adequate. During the last 10 years growing emphasis has been placed on dating of older archaeological materials such as flint, stones, and calcite. Finally, within the last 2 years there has been a 10-fold expansion in the application of TL to the dating of unheated sediments.

REFERENCES

1. **Harvey, E. N.**, *A History of Luminescence from Earliest Times until 1900*, American Philosophical Society, Philadelphia, 1957.
2. **Trenkle, W.**, Über Luminescenzerscheinungen, *Berichte des Naturwissenschaftlich*, Vereines Regensburg, X Heft. für die Jahre 1903 und 1904, 1905.
3. **Hegemann, F. and Steinmetz, H.**, Über die Thermoluminescenz in Ihrer Minerogenetischen Bedeutung, *Zentralbl. Mineral.*, 24, 1933.
4. **Köhler, A. and Leitmeier, H.**, Die naturliche Thermoluminescenz bei Mineralien und Gesteinen, *Z. Kristallogr.*, 87, 146, 1934.
5. **Daniels, F., Boyd, C. A., and Saunders, D. F.**, Thermoluminescence as a research tool, *Science*, 117, 343, 1953.
6. **Grögler, N., Houtermans, F. G., and Stauffer, H.**, Über die Datierung von Keramik und Ziegel durch Thermoluminescenz, *Helv. Phys. Acta*, 33, 95, 1960.
7. **Aitken, M. J., Tite, M. S., and Reid, J.**, Thermoluminescent dating of ancient pottery, *Nature (London)*, 202, 1032, 1964.
8. **Mazess, R. B. and Zimmerman, D. W.**, Pottery dating from thermoluminescence, *Science*, 152, 347, 1966.
9. **Ralph, E. K. and Han, M. C.**, Dating of pottery by thermoluminescence, *Nature (London)*, 210, 245, 1966.
10. **McDougall, D. J., Ed.**, *Thermoluminescence of Geological Materials*, Academic Press, New York, 1968.
11. **Aitken, M. J.**, *Physics and Archaeology*, Oxford University Press, Oxford, 1974.
12. **Aitken, M. J.**, Thermoluminescence and the archaeologist, *Antiquity*, 51, 11, 1977.
13. **Seeley, M.-A.**, Thermoluminescent dating in its application to archaeology: a review, *J. Archaeol. Sci.*, 2, 17, 1975.
14. **McDougall, D. J.**, Natural thermoluminescence of igneous rocks and associated ore deposits, in *Thermoluminescence of Geological Materials*, McDougall, D. J., Ed., Academic Press, New York, 1968, 527.
15. **Siegel, F. R., Vaz, J. E., and Ronca, L. B.**, Thermoluminescence of clay minerals, in *Thermoluminescence of Geological Materials*, McDougall, D. J., Ed., Academic Press, New York, 1968, 635.
16. **Nishita, H., Hamilton, M., and Haug, R. M.**, Natural thermoluminescence of soils, minerals, and certain rocks, *Soil Sci.*, 117, 211, 1974.
17. **Sankaran, A. V., Nambi, K. S. V., and Sunta, C. M.**, Current Status of Thermoluminescence Studies on Minerals and Rocks, No. 1156, Bhabha Atomic Research Centre, Bombay, India, 1982.
18. **Huntley, D. J. and Bailey, D. C.**, Obsidian source identification by thermoluminescence, *Archaeometry*, 20, 159, 1978.
19. **Charlet, J. M.**, Utilisation des courbes de thermoluminescence artificielle dans l'étude des séries sédimentaires détritiques, *Bull. Soc. Géol. Fr.*, 11, 287, 1969.
20. **Wintle, A. G.**, Thermoluminescence dating of minerals — traps for the unwary, *J. Electrostatics*, 3, 281, 1977.
21. **Fleming, S. J.**, The pre-dose technique: a new thermoluminescent dating method, *Archaeometry*, 15, 13, 1973.
22. **Aitken, M. J. and Fleming, S. J.**, Thermoluminescence dosimetry in archaeological dating, in *Topics in Radiation Dosimetry, Suppl. 1*, Attix, F. H., Ed., Academic Press, New York, 1972, 1.
23. **Wintle, A. G.**, Thermal quenching of thermoluminescence in quartz, *Geophys. J. R. Astron. Soc.*, 41, 107, 1975.
24. **Hoogenstraaten, W.**, Electron traps in zinc-sulphide phosphors, *Philips Res. Rep.*, 13, 515, 1958.
25. **Pasternack, E. S., Gaines, A. M., and Levy, P. W.**, Second order kinetics in the thermoluminescence of natural $NaAlSi_3O_8$ (albite), *Bull. Am. Phys. Soc.*, 22, 409, 1977.
26. **Levy, P. W.**, Thermoluminescence studies having applications to geology and archaeometry, *PACT*, 3, 466, 1979.
27. **Levy, P. W.**, Thermoluminescence and optical bleaching in minerals exhibiting second order kinetics and other charge retrapping characteristics, *PACT*, 6, 224, 1982.

28. **Aitken, M. J.**, Thermoluminescent dating in archaeology: introductory review, in *Thermoluminescence of Geological Materials*, McDougall, D. J., Ed., Academic Press, New York, 1968, 369.

29. **Gunn, N. M. and Murray, A. S.**, Geomagnetic field magnitude variations in Peru derived from archaeological ceramics dated by thermoluminescence, *Geophys. J. R. Astron. Soc.*, 62, 345, 1980.

30. **Mejdahl, V. and Winther-Nielsen, M.**, TL dating based on feldspar inclusions, *PACT*, 6, 426, 1982.

31. **Aitken, M. J., Fleming, S. J., Reid, J., and Tite, M. S.**, Elimination of spurious thermoluminescence, in *Thermoluminescence of Geological Materials*, McDougall, D. J., Ed., Academic Press, New York, 1968, 133.

32. **Wintle, A. G.**, Detailed study of a thermoluminescent mineral exhibiting anomalous fading, *J. Luminescence*, 15, 385, 1977.

33. **Wintle, A. G.**, Anomalous fading of thermoluminescence in mineral samples, *Nature (London)*, 245, 143, 1973.

34. **Garlick, G. F. J. and Robinson, I.**, The thermoluminescence of lunar samples, in *The Moon*, Runcorn, S. K. and Urey, H. C., Eds., Symposium IAU, Reidel, Dordrecht, The Netherlands, 1972, 324.

35. **Warren, S. E.**, Thermoluminescence dating: an assessment of the dose-rate from rubidium, *Archaeometry*, 20, 71, 1978.

36. **Bell, W. T.**, The assessment of the radiation dose-rate for thermoluminescence dating, *Archaeometry*, 18, 107, 1976.

37. **Bell, W. T.**, Thermoluminescence dating: revised dose-rate data, *Archaeometry*, 19, 99, 1977.

38. **Bell, W. T.**, Thermoluminescence dating: radiation dose-rate data, *Archaeometry*, 21, 243, 1979.

39. **Zimmerman, D. W.**, Relative thermoluminescence effects of alpha- and beta-radiation, *Radiat. Eff.*, 14, 81, 1972.

40. **Desai, V. S. and Aitken, M. J.**, Radon escape from pottery: effect of wetness, *Archaeometry*, 16, 95, 1974.

41. **Mejdahl, V.**, Thermoluminescence dating: beta-dose attenuation in quartz grains, *Archaeometry*, 21, 61, 1979.

42. **Aitken, M. J. and Bowman, S. G. E.**, Thermoluminescent dating: assessment of alpha particle contribution, *Archaeometry*, 17, 132, 1975.

43. **Ockerman, J. B. and Daniels, F.**, α-Radioactivity of some rocks and common materials, *J. Phys. Chem.*, 58, 926, 1954.

44. **Fisenne, I. M. and Keller, H. W.**, A short history of ZnS on mylar as an α-scintillation detector, *Health Phys.*, 40, 739, 1981.

45. **Masters, B. J.**, Experimental studies of ZnS α-particle counters and methods for minimizing detector background, in Proc. 18th Int. Reliability Phys. Symp., 1980, 269.

46. **Sasidharan, R., Sunta, C. M., and Nambi, K. S. V.**, TL dating: error implications in case of undetermined U-Th concentration ratio in pottery samples, *Ancient TL*, 2, 8, 1978.

47. **Turner, R. C., Radley, J. M., and Mayneord, W. V.**, The alpha-ray activity of human tissues, *Br. J. Radiol.*, 31, 397, 1958.

48. **Giffin, C., Kaufman, A., and Broecker, W.**, Delayed coincidence counter for the assay of actinon and thoron, *J. Geophys. Res.*, 68, 1749, 1963.

49. **Huntley, D. J. and Wintle, A. G.**, The use of alpha scintillation counting for measuring Th-230 and Pa-231 contents of ocean sediments, *Can. J. Earth Sci.*, 18, 419, 1981.

50. **Aitken, M. J.**, Radon loss evaluation by alpha counting, *PACT*, 2, 104, 1978.

51. **Pernicka, E. and Wagner, G. A.**, Radioactive equilibrium and dose-rate determination in TL dating, *PACT*, 6, 132, 1982.

52. **Murray, A. S.**, Discrepancies in the alpha counting of known activity samples, *PACT*, 6, 53, 1982.

53. **Prescott, J. R. and Jensen, H. E.**, Low-level thorium and uranium determination for thermoluminescent dating, *Austr. At. En.*, 23, 23, 1980.

54. **Bowman, S. G. E.**, Alpha particle ranges in pottery, *PACT*, 6, 61, 1982.

55. **Zimmerman, D. W.**, Uranium distributions in archaeological ceramics: dating of radioactive inclusions, *Science*, 174, 818, 1971.

56. **Malik, S. R., Durrani, S. A., and Fremlin, J. H.**, A comparative study of the spatial distribution of uranium and of TL-producing minerals in archaeological materials, *Archaeometry*, 15, 249, 1973.

57. **Walton, A. J. and Debenham, N. C.**, Spatial distribution studies of thermoluminescence using a high-gain image intensifier, *Nature (London)*, 284, 42, 1980.

58. **Bowman, S. G. E. and Seeley, M.-A.**, The British museum flint dating project, *PACT*, 2, 151, 1978.

59. **Gillot, P. Y., Valladas, G., and Reyss, J. L.**, Dating of lava flow using a granitic enclave, application to the Laschamp magnetic event, *PACT*, 2, 165, 1978.

60. **Sutton, S. R. and Zimmerman, D. W.**, Thermoluminescent dating using zircon grains from archaeological ceramics, *Archaeometry*, 18, 125, 1976.

61. **Wagner, G. A.**, Dose rate evaluation for thermoluminescence dating by fission track counting, in Proc. 1976 Symp. Archaeometry and Archaeological Prospection, National Museum of Antiquities of Scotland, Edinburgh, 1980, 393.

62. **Filberth, E. W., Rowe, M. W., and Shaefer, H. J.**, Uranium in Hueco and Guadaloupe Mountain Indian ceramics, *Archaeometry*, 22, 107, 1980.

63. **Meakins, R. L., Dickson, B. L., and Kelly, J. C.**, Gamma ray analysis of K, U and Th, for dose-rate estimation in thermoluminescent dating, *Archaeometry*, 21, 79, 1979.

64. **Nielsen, S.**, *In situ* Measurements of Environmental Gamma Radiation Using a Mobile Ge(Li) Spectrometer System, Rep. No. 367, Risø National Laboratory, Risø, Denmark, 1977.

65. **Murray, A. S. and Aitken, M. J.**, The occurrence and importance of radioactive disequilibria in TL samples, *PACT*, 6, 155, 1982.

66. **Zimmerman, D. W.**, Thermoluminescent dating using fine grains from pottery, *Archaeometry*, 13, 29, 1971.

67. **Fleming, S. J.**, Thermoluminescent dating: refinement of the quartz inclusion method, *Archaeometry*, 12, 133, 1970.

68. **Fleming, S. J.**, *Thermoluminescence Techniques in Archaeology*, Clarendon Press, Oxford, 1979.

69. **Bell, W. T.**, Attenuation factors for the absorbed radiation dose in quartz inclusions for thermoluminescence dating, *Ancient TL*, 8, 2, 1979.

70. **Bell, W. T.**, Alpha dose attenuation in quartz grains, *Ancient TL*, 12, 4, 1980.

71. **Wintle, A. G. and Huntley, D. J.**, Thermoluminescence dating of a deep-sea sediment core, *Nature (London)*, 279, 710, 1979.

72. **Wintle, A. G.**, A thermoluminescence dating study of some Quaternary calcites: potential and problems, *Can. J. Earth Sci.*, 15, 1977, 1978.

73. **Sutton, S. R. and Zimmerman, D. W.**, Thermoluminescence dating: radioactivity in quartz, *Archaeometry*, 20, 66, 1978.

74. **Bell, W. T. and Zimmerman, D. W.**, The effect of HF etching on the morphology of quartz inclusions for thermoluminescence dating, *Archaeometry*, 20, 63, 1978.

75. **Valladas, H. and Valladas, G.**, Effet de l'irradiation a sur des grains de quartz, *PACT*, 6, 171, 1982.

76. **Guérin, G. and Valladas, G.**, Thermoluminescence dating of volcanic plagioclases, *Nature (London)*, 286, 697, 1980.

77. **Guérin, G.**, Évaluation des débits de dose annuels utilisés pour la datation des coulées volcaniques, *PACT*, 6, 179, 1982.

78. **Mejdahl, V., Bell, W. T., and Winther-Nielsen, M.**, Datering af keramik fra arkæologiske udgravninger ved hjælp af termoluminescens (TL), *Aarbøger for Nordisk Oldkyndighed og Historie 1979*, 122, 1980.

79. **Mejdahl, V.**, An automated procedure for thermoluminescence dating of pottery and burnt stones, *PACT*, 7, 83, 1982.

80. **Aitken, M. J.**, Dose-rate evaluation, *PACT*, 2, 18, 1978.

81. **Fleming, S. J.**, Development and application of calcium fluoride for evaluation of dosage with a radioactive powder matrix, in Proc. 2nd Int. Conf. Luminescence Dosimetry, U.S. A.E.C.-CONF-680920, NTIS, Springfield, Va., 1968, 465.

82. **Mejdahl, V.**, Feldspar inclusion dating of burnt granitic stones, *PACT*, 9, 351, 1983.

83. **McKerrell, H. and Mejdahl, V.**, Progress and problems with automated TL dating, Proc. 16th Int. Symp. Archaeometry and Archaeological Prospection, Risø-M-2265, Risø National Laboratory, Roskilde, 1981.

84. **Liritzis, Y. and Galloway, R. B.**, A new approach to the beta dosimetry of ceramics for thermoluminescence dating, *Nuclear Instruments and Methods in Physics Research*, 201, 503, 1982.

85. **Bailiff, I. K. and Aitken, M. J.**, Use of thermoluminescence dosimetry for evaluation of internal beta dose-rate in archaeological dating, *Nucl. Instru. Methods*, 173, 423, 1980.

86. **Bailiff, I. K.**, Beta-TLD apparatus for small samples, *PACT*, 6, 72, 1982.

87. **Majdahl, V.**, Thermoluminescence dating: a thermoluminescence technique for beta-ray dosimetry, *PACT*, 2, 35, 1978.

88. **Mejdahl, V.**, Thermoluminescence dating based on quartz and feldspar inclusions, *PACT Educ.*, 2, in press.

89. **Bøtter-Jensen, L. and Beckmann, P.**, A versatile automatic sample changer for reading of thermoluminescence dosimeters and phosphors, in Proc. 2nd Int. Conf. Luminescence Dosimetry, U.S. A.E.C.-CONF-680920, NTIS, Springfield, Va., 1968, 640.

90. **Bøtter-Jensen, L.**, Read-out instruments for solid thermoluminescence dosimeters using hot nitrogen as the heating medium, in *Advances in Physical and Biological Radiation Dosimeters*, IAEA, Vienna, 1971, 113.

91. **Zimmerman, D. W.**, Introduction to basic procedures for sample preparation and thermoluminescence measurement of ceramics, *PACT*, 2, 1, 1978.

92. **Mejdahl, V. and Winther-Nielsen, M.**, Dating of Danish ceramics by means of the quartz inclusion technique, *PACT*, 2, 131, 1978.

93. **Audric, T. and Bouqier, L.,** Collapsing behaviour of some loesses and soils, *Q. J. Eng. Geol.*, 9, 265, 1976.

94. **Berger, M. J.,** Distribution of absorbed dose around point sources of electrons and beta particles in water and other media, *J. Nucl. Med.*, Suppl. No. 5, MIRD/Pamphlet No. 7, 5, 1971.

95. **Berger, M. J.,** Improved Point Kernels for Electron and Beta-Ray Dosimetry, NBSIR, National Bureau of Standards, Washington, D.C., 1973, 73.

96. **Cross, W. G.,** Variation of beta dose attenuation in different media, *Phys. Med. Biol.*, 13, 611, 1968.

97. **Wu, S. and Kendall, F. H.,** Radon emanation from Hong Kong soils and sherds, *PACT*, 6, 34, 1982.

98. **Carriveau, G. W. and Harbottle, G.,** Radon and thoron emanation measurements and the effect of ground water, *PACT*, 6, 145, 1982.

99. **Faure, G.,** *Principles of Isotope Geology*, John Wiley & Sons, New York, 1977, chap. 16.

100. **Hedges, R. E. M. and McLellan, M.,** On the cation exchange capacity of fired clay and its effect on the chemical and radiometric analysis of pottery, *Archaeometry*, 18, 203, 1976.

101. **Murray, A. S.,** Studies of the stability of radioisotope concentrations and their dependence on grain size, *PACT*, 6, 216, 1982.

102. **Clark, R. M.,** A calibration curve for radiocarbon dates, *Antiquity*, XLIX, 251, 1975.

103. **Olsson, I. U.,** On the work of the International Committee on Calibration of the Radiocarbon Dating Time Scale, *PACT*, 7, 37, 1982.

104. **Prescott, J. R. and Stephan, L.,** The contribution of cosmic radiation to the environmental dose for thermoluminescent dating, *PACT*, 6, 17, 1982.

105. **Aitken, M. J.,** Thermoluminescent dosimetry of environmental radiation on archaeological sites, *Archaeometry*, 11, 109, 1969.

106. **Mejdahl, V.,** Measurement of environmental radiation at archaeological sites by means of TL dosimeters, *PACT*, 2, 70, 1978.

107. **Mejdahl, V.,** Measurement of environmental radiation at archaeological excavation sites, *Archaeometry*, 12, 147, 1970.

108. **Murray, A. S., Bowman, S. G. E., and Aitken, M. J.,** Evaluation of the gamma dose-rate contribution, *PACT*, 2, 84, 1978.

109. **Mejdahl, V.,** Dosimetry techniques in thermoluminescence dating, Rep. No. 261, Risø National Laboratory, Roskilde, 1972.

110. **Mehta, S. K. and Sengupta, S.,** Gamma dosimetry with Al_2O_3 thermoluminescent phosphor, *Phys. Med. Biol.*, 21, 955, 1976.

111. **Løvborg, L. and Kirkegaard, P.,** Response of 3'' × 3'' NaI(Tl) detectors to terrestrial gamma radiation, *Nucl. Instrum. Methods*, 121, 239, 1974.

112. **Bell, W. T., Mejdahl, V., and Winther-Nielsen, M.,** The refinement of the automated TL dating procedure and perspectives for the archaeological application of the method as demonstrated by the results from sites of known age, *Rev. Archéométrie*, 4, 127, 1980.

113. **Liritzis, Y. and Galloway, R. B.,** A new technique for calibrating a NaI(Tl) scintillometer used to measure gamma dose-rates in archaeological sites, *Nucl. Instrum. Methods*, 174, 593, 1980.

114. **Brownell, G. L., Ellett, W. H., and Reddy, A. R.,** Absorbed fractions for photon dosimetry, *J. Nucl. Med.*, Suppl. No. 1, MIRD/Pamphlet No. 3, 27, 1968.

115. **Ellett, W. H. and Humes, R. M.,** Absorbed fraction for small volumes containing photon-emitting radioactivity, *J. Nucl. Med.*, Suppl. No. 5, MIRD/Pamphlet No. 8, 25, 1971.

116. National Council on Radiation Protection and Measurements, Environmental Radiation Measurements, Rep. No. 50, NCRP, Washington, D.C., 1977.

117. **Tanner, A. B.,** Radon migration in the ground: a review, in *The Natural Radiation Environment*, Adams, J. A. S. and Lowder, W. M., Eds., University of Chicago Press, Chicago, 1964, 161.

118. **Liritzis, Y. and Galloway, R. B.,** Correlation of variations in the gamma-ray dose-rate in soil with meteorological factors, *Archaeometry*, 23, 109, 1981.

119. **Fleming, S. J.,** Supralinearity corrections in fine-grain thermoluminescence dating: a reappraisal, *Archaeometry*, 17, 122, 1975.

120. **Bowman, S. G. E.,** Dependence of supralinearity on pre-dose: some observations, *Archaeometry*, 17, 129, 1975.

121. **Chen, R. and Bowman, S. G. E.,** Superlinear growth of thermoluminescence due to competition during irradiation, *PACT*, 2, 216, 1978.

122. **Bell, W. T.,** Thermoluminescence dates for the Lake Mungo aboriginal fireplaces and the consequences for radiocarbon dating, submitted.

123. **Huxtable, J. and Aitken, M. J.,** Thermoluminescent dating of Lake Mungo geomagnetic polarity excursion, *Nature (London)*, 265, 40, 1977.

124. **Valladas, G. and Gillot, P. Y.,** Dating of the Olby lava flow using heated quartz pebbles: some problems, *PACT*, 2, 141, 1978.

125. **Guérin, G.,** Croissance de la thermoluminescence en fonction de la dose des feldspaths d'origine volcanique, *PACT*, 6, 417, 1982.

126. **Bailiff, I. K.,** Use of phototransfer for the anomalous fading of thermoluminescence, *Nature (London)*, 264, 531, 1976.

127. **Bailiff, I. K., Bowman, S. G. E., Mobbs, S. F., and Aitken, M. J.,** The phototransfer technique and its use in thermoluminescence dating, *J. Electrostatics*, 3, 269, 1977.

128. **Bowman, S. G. E.,** Phototransferred thermoluminescence in quartz and its potential use in dating, *PACT*, 3, 381, 1979.

129. **Sasidharan, R., Sunta, C. M., and Nambi, K. S. V.,** Phototransfer method of determining archaeological dose of pottery sherds, *PACT*, 3, 401, 1979.

130. **Mobbs, S. F.,** Phototransfer at low temperatures, *PACT*, 3, 407, 1979.

131. **Kristianpoller, N. and Aitken, M. J.,** Low temperature phototransfer studies, *PACT*, 6, 473, 1982.

132. **Fleming, S. J.,** Thermoluminescence authenticity testing of ancient ceramics using radiation-sensitivity changes in quartz, *Naturwissenschaften*, 59, 145, 1972.

133. **Fleming, S. J. and Thompson, J.,** Quartz as a heat-resistant dosimeter, *Health Phys.*, 18, 567, 1970.

134. **Zimmerman, J.,** The radiation-induced increase of the 100°C thermoluminescence sensitivity of fired quartz, *J. Phys. C*, 4, 3265, 1971.

135. **Aitken, M. J.,** Pre-dose dating: predictions from the model, *PACT*, 3, 319, 1979.

136. **Wright, D. A.,** Predose TL measurements at Durham, *PACT*, 3, 336, 1979.

137. **Chen, R.,** Saturation of sensitization of the 110°C TL peak in quartz and its potential application in the pre-dose technique, *PACT*, 3, 325, 1979.

138. **Bailiff, I. K.,** Dating of pottery from a Saxon monastery by the pre-dose technique, *PACT*, 6, 468, 1982.

139. **Bailiff, I. K.,** Pre-dose dating: high-S_o sherds, *PACT*, 3, 345, 1979.

140. **Wintle, A. G. and Huntley, D. J.,** Thermoluminescence dating of sediments, *Quat. Sci. Rev.*, 1, 31, 1982.

141. **Wintle, A. G.,** Thermoluminescence properties of fine grain minerals in loess, *Soil Sci.*, 134, 164, 1982.

142. **Wintle, A. G. and Huntley, D. J.,** Thermoluminescence dating of ocean sediments, *Can. J. Earth Sci.*, 17, 348, 1980.

143. **Wintle, A. G.,** Thermoluminescence dating of late Devensian loesses in southern England, *Nature (London)*, 289, 479, 1981.

144. **Berger, G. W. and Huntley, D. J.,** Thermoluminescence dating of terrigenous sediments, *PACT*, 6, 495, 1982.

145. **Singhvi, A. K., Sharma, Y. P., and Agrawal, D. P.,** Thermoluminescence dating of sand dunes in Rajasthan, India, *Nature (London)*, 295, 313, 1982.

146. **Prószyńska, H.,** TL dating of some subaerial sediments from Poland, *PACT*, 9, 539, 1983.

147. **Huxtable, J.,** Fine grain dating, *PACT*, 2, 7, 1978.

148. **Singhvi, A. K. and Zimmerman, D. W.,** The luminescent minerals in fine-grain samples from archaeological ceramics, *PACT*, 2, 12, 1978.

149. **Fleming, S. J.,** The quartz inclusion method, *PACT*, 2, 125, 1978.

150. **Bell, W. T. and Mejdahl, V.,** Beta source calibration and its dependency on grain transparency, *Archaeometry*, 23, 231, 1981.

151. **Andres, U.,** *Magnetohydrodynamic and Magnetostatic Methods of Mineral Separation*, John Wiley & Sons, New York, 1976.

152. **Janér, J.-H. and Jungner, H.,** A simple method for separation of quartz in TL dating, *PACT*, 6, 214, 1982.

153. **Hütt, G., Mangerud, J., and Punning, J.-M.,** Thermoluminescence dating of Eemian-early Weichselian sequence at Fjøsanger, Western Norway, *PACT*, 9, 593, 1983.

154. **Kronborg, C.,** Preliminary results from age determinations by TL of interglacial and interstadial waterlaid sediments, *PACT*, 9, 595, 1983.

155. **Fleming, S. J. and Stoneham, D.,** The subtraction technique of thermoluminescent dating, *Archaeometry*, 15, 229, 1973.

156. **Plachy, A. L. and Sutton, S. R.,** Determination of the dose-rate o quartz in granite, *PACT*, 6, 188, 1982.

157. **Zimmerman, D. W., Yuhas, M. P., and Meyers, P.,** Thermoluminescence authenticity measurements on core material from the bronze horse of the New York Metropolitan Museum of Art, *Archaeometry*, 16, 19, 1974.

158. **Langouet, L., Roman, A., and Gonzales, R.,** Datation de poteries anciennes par la méthode D.A.T.E., 1976 Symp. on Archaeometry and Archaeological Prospection, National Museum of Antiquities of Scotland, Edinburgh, 1980, 312.

159. **Langouet, L., Roman, A., Deza, A., Brito, O., Concha, G., and Asenjo de Roman, C.,** Datation relative par thermoluminescence, Méthode DATE, *Rev. Archéometrie*, 3, 57, 1979.

160. **Charalambous, S. and Michael, C.,** A new method of dating pottery by thermoluminescence, *Nucl. Instrum. Methods*, 137, 565, 1976.

161. **Charalambous, S., Hasan, F., Michael, C., Siona, A., and Tzamarias, S.,** Dating using the shape of the TL glow curve, *PACT,* 6, 265, 1982.

162. **Schvoerer, M., Bechtel, F., Deshouillers, J.-M., Dautant, A., and Gallois, B.,** Datation par gamma thermoluminescence: recherches sur une nouvelle méthode, *PACT,* 6, 86, 1982.

163. **Aitken, M. J.,** Interlaboratory calibration of alpha and beta sources, *PACT,* 3, 443, 1979.

164. **Pernicka, E. and Wagner, G. A.,** Primary and interlaboratory calibration of beta sources using quartz as thermoluminescent phosphor, *Ancient TL,* 6, 2, 1979.

165. **Wagner, G. A. and Pernicka, E.,** Beta source calibration by using quartz: a re-appraisal, *PACT,* 6, 515, 1982.

166. **Kalefezra, J. and Horowitz, Y. S.,** Electron backscattering corrections for beta dose-rate estimations in archaeological objects, *PACT,* 3, 428, 1979.

167. **Murray, A. S. and Wintle, A. G.,** Beta source calibration, *PACT,* 3, 419, 1979.

168. **Wintle, A. G. and Aitken, M. J.,** Absorbed dose from a beta source as shown by thermoluminescence dosimetry, *Int. J. Appl. Radiat. Isot.,* 28, 625, 1977.

169. **Wintle, A. G. and Murray, A. S.,** Thermoluminescence dating: reassessment of the fine grain dose rate, *Archaeometry,* 19, 95, 1977.

170. **Wintle, A. G.,** Thermoluminescence dating: a review of recent applications to non-pottery materials, *Archaeometry,* 22, 113, 1980.

171. **Wagner, G. A.,** Application of TLD for dating: a review, in *Applied Thermoluminescence Dosimetry,* Oberhofer, M. and Scharmann, A., Eds., Adam Hilger Ltd., Bristol, 1981, 347.

172. **Aitken, M. J. and Alldred, J. C.,** The assessment of error limits in thermoluminescent dating, *Archaeometry,* 14, 257, 1972.

173. **Aitken, M. J.,** Thermoluminescent age evaluation and assessment of error limits, *Archaeometry,* 18, 233, 1976.

174. **Wagner, G. A.,** Thermoluminineszenz-Datierungen am Töpferofen Koberg 15 in Lübeck, *Lübecker Schriften zur Archaeologie und Kulturgeschichte (LSAK),* 3, 83, 1980.

175. **Agrawal, D. P., Bhandari, N., Lal, B. B., and Singhvi, A. K.,** Thermoluminescence dating of pottery from Sringaverapura — a Ramayana site, *Proc. Indian Acad. Sci. (Earth Planet. Sci.),* 90, 161, 1981.

176. **Valladas, H.,** Datation par thermoluminescence de grés brûlés de foyers de quatre gisements du Magdalénien Final du Bassin Parisien, *C. R. Acad. Seances,* 292, 355, 1981.

177. **Bechtel, F., Schvoerer, M., Rouanet, J.-F., and Gallois, B.,** Extension à la préhistoire, à l'océanographie et à la volcanologie de la méthode de datation par thermoluminescence, *PACT,* 3, 481, 1979.

178. **Ichikawa, Y. and Nagatomo, T.,** Thermoluminescence dating of burnt sandstones from Senpukuji cave, *PACT,* 2, 174, 1978.

179. **Bowman, S. G. E.,** Thermoluminescence studies on burnt flint, *PACT,* 6, 353, 1982.

180. **Aitken, M. J. and Wintle, A. G.,** Thermoluminescence dating of calcite and burnt flint, *Archaeometry,* 19, 100, 1977.

181. **Wintle, A. G. and Aitken, M. J.,** Thermoluminescence of burnt flint: application to a lower palaeolithic site, Terra Amata, *Archaeometry,* 19, 111, 1977.

182. **Valladas, H.,** Thermoluminescence dating of burnt stones from prehistoric site, *PACT,* 2, 180, 1978.

183. **Danon, J., Enriquez, C. R., Zuleta, E., Beltrão, M. M. C., and Poupeau, G.,** Thermoluminescence dating of archaeologically heated cherts. A case study: the Alice Boër site, *PACT,* 6, 370, 1982.

184. **Huxtable, J.,** Fine grain thermoluminescence (TL) techniques applied to flint dating, *PACT,* 6, 346, 1982.

185. **Huxtable, J. and Jacobi, R. M.,** Thermoluminescence dating of burnt flints from a British mesolithic site: Longmoor Inclosure, East Hampshire, *Archaeometry,* 24, 164, 1982.

186. **Bowman, S. G. E. and Sieveking, G. de G.,** Thermoluminescence dating of burnt flint from Combe Grenal, *PACT,* 9, 253, 1983.

187. **Sabels, B. E.,** Age studies on basaltic lava flows using natural radioactivity and thermoluminescence, in *Radioactive Dating,* IAEA, Vienna, 1962, 87.

188. **Aitken, M. J., Fleming, S. J., Doell, R. R., and Tanguy, J. C.,** Thermoluminescent study of lavas from Mount Etna and other historic flows: preliminary results, in *Thermoluminescence of Geological Materials,* McDougall, D. J., Ed., Academic Press, New York, 1968, 359.

189. **Hwang, F. S. W.,** Thermoluminescence dating applied to volcanic lava, *Nature (London),* 227, 940, 1970.

190. **Berry, A. L.,** Thermoluminescence of Hawaiian basalts, *J. Geophys. Res.,* 78, 6863, 1973.

191. **May, R. J.,** Thermoluminescence dating of Hawaiian alkali basalts, *J. Geophys. Res.,* 82, 3023, 1977.

192. **May, R. J.,** Thermoluminescence dating of Hawaiian basalt, *U.S. Geol. Surv. Prof. Pap.,* 1095, 1979.

193. **Huxtable, J., Aitken, M. J., and Bonhommet, N.,** Thermoluminescence dating of sediment baked by lava flows of the Chaîne des Puys, *Nature (London),* 275, 207, 1978.

194. **Pei, J.-X.,** Thermoluminescence dating of the cave-deposits in Zhoukoudian and the volcanics of baked layer in Datong, Shanxi Province, *Sci. Geol. Sin.,* 10, 403, 1980.

195. **Brou, R. and Valladas, G.,** Appareil pour la mesure de la thermoluminescence de petits échantillons, *Nucl. Instrum. Methods,* 127, 109, 1975.

196. **Valladas, G. and Valladas, H.**, High temperature thermoluminescence, *Archaeo-Physica*, 12, 506, 1978.
197. **Guérin, G., Gillot, P. Y., Le Garrec, M.-J., and Brousse, R.**, Age subactuel des dernières manifestations éruptives du Mont-Dore et du Cézallier, *C. R. Acad. Séances*, 292, 855, 1981.
198. **Bothner, M. H. and Johnson, N. M.**, Natural thermoluminescent dosimetry in late Pleistocene pelagic sediments, *J. Geophys. Res.*, 74, 5331, 1969.
199. **Huntley, D. J. and Johnson, H. P.**, Thermoluminescence as a potential means of dating siliceous ocean sediments, *Can. J. Earth Sci.*, 13, 593, 1976.
200. **Sadasivan, S., Nambi, K. S. V., and Murali, A. V.**, Geochemical and thermoluminescence studies of the shales from the off-shore drill core, west coast of India, *Mod. Geol.*, 8, 13, 1981.
201. **Wintle, A. G.**, Thermoluminescence dating of loess, *PACT*, 6, 486, 1982.
202. **Wintle, A. G. and Brunnacker, K.**, Age of volcanic tuff in Rheinhessen obtained by thermoluminescence dating of loess, *Naturwissenschaften*, 69, 181, 1982.
203. **Hütt, G. and Smirnov, A.**, Thermoluminescence dating in the Soviet Union, *PACT*, 7, 97, 1982.
204. **Dreimanis, A., Hütt, G., Raukas, A., and Whippey, P. W.**, Dating methods of pleistocene deposits and their problems. I. Thermoluminescence dating, *Geosci. Can.*, 5, 55, 1978.
205. **Hütt, G., Smirnov, A., and Tale, I.**, On the application of thermoluminescence of natural quartz to the study of geochronology of sedimentary deposits, *PACT*, 3, 362, 1979.
206. **Hütt, G., Punning, J.-M., and Smirnov, A.**, The potential use of the TL dating method in chronological studies of late Pleistocene, *PACT*, 7, 105, 1982.
207. **Troitsky, L., Punning, J.-M., Hütt, G., and Rajamäe, R.**, Pleistocene glaciation chronology of Spitsbergen, *Boreas*, 8, 401, 1979.
208. **Sears, D. W.**, Thermoluminescence of meteorites: relationships with their K-Ar age and their shock and reheating history, *Icarus*, 44, 190, 1980.
209. **Liener, A. and Geiss, J.**, Thermoluminescence measurements on chondritic meteorites, in *Thermoluminescence of Geological Materials*, McDougall, D. J., Ed., Academic Press, New York, 1968, 559.
210. **Melcher, C. L.**, Thermoluminescence of meteorites and their orbits, *Earth Planet. Sci. Lett.*, 52, 39, 1981.
211. **Vaz, J. E.**, Lost City meteorite: determination of the temperature gradient induced by atmospheric friction using TL, *Meteoritics*, 6, 207, 1971.
212. **Durrani, S. A.**, Use of thermoluminescence for meteorite dating, *PACT*, 6, 384, 1982.
213. **Melcher, C. L.**, Thermoluminescence of meteorites and their terrestrial ages, *Geochim. Cosmochim. Acta*, 45, 615, 1981.
214. **McKeever, S. W. S.**, Dating of meteorite falls using thermoluminescence: application to Antarctic meteorites, *Earth Planet. Sci. Lett.*, 58, 419, 1982.
215. **Ewbank, W. B.**, Status of transactinium nuclear data in the evaluated nuclear structure data file, in Transactinium Isotope Nuclear Data-1979, IAEA-TECDOC-232, IAEA, Vienna, 1980, 109.
216. **Valladas, G.**, personal communication.
217. **Aitken, M. J.**, *PACT*, 6, 69, 1983.

INDEX

M

N

O

P

Q

R